general editor John M. MacKenzie

When the 'Studies in Imperialism' series was founded more than twenty years ago, emphasis was laid upon the conviction that 'imperialism as a cultural phenomenon had as significant an effect on the dominant as on the subordinate societies'. With more than sixty books published, this remains the prime concern of the series. Cross-disciplinary work has indeed appeared covering the full spectrum of cultural phenomena, as well as examing aspects of gender and sex, frontiers and law, science and the environment, language and literature, migration and patriotic societies, and much else. Moreover, the series has always wished to present comparative work on European and American imperialism, and particularly welcomes the submission of books in these areas. The fascination with imperialism, in all its aspects, shows no sign of abating, and this series will continue to lead the way in encouraging the widest possible range of studies in the field. 'Studies in Imperialism' is fully organic in its development, always seeking to be at the cutting edge, responding to the latest interests of scholars and the needs of this ever-expanding area of scholarship.

Race and empire

Manchester University Press

Race and empire

EUGENICS IN COLONIAL KENYA

Chloe Campbell

MANCHESTER UNIVERSITY PRESS
Manchester and New York

distributed exclusively in the USA by PALGRAVE

Published by Manchester University Press
Oxford Road, Manchester M13 9NR, UK
and Room 400, 175 Fifth Avenue, New York, NY 10010, USA
www.manchesteruniversitypress.co.uk

Distributed in the United States exclusively by
Palgrave Macmillan, 175 Fifth Avenue,
New York, NY 10010, USA

Distributed in Canada exclusively by
UBC Press, University of British Columbia, 2029 West Mall,
Vancouver, BC, Canada V6T 1Z2

British Library Cataloguing-in-Publication Data is available

Library of Congress Cataloging-in-Publication Data is available

ISBN 978 0 7190 7161 4 paperback

First published by Manchester University Press in hardback 2007

This paperback edition first published 2012

Printed by Lightning Source

CONTENTS

ACKNOWLEDGMENTS

First of all, thanks must go to David Anderson, who has been unerring in both his encouragement and perceptive criticism throughout this project.

In Nairobi, I was lucky to have the help and support provided by all at the British Institute, by Godfrey Muriuki and Shirin Walji. Thanks are also due to all at the Kenya National Archives, in particular Musila Musembe and Peterson Kithuka Nthiwa. Mrs Owles, Peter Paterson and Laurie Slade showed warmth and welcoming interest in my subject.

Funding from the AHRB made it possible for me to undertake the research that forms the basis of this book and I gratefully acknowledge this support. I am also grateful to the Galton Institute, London, for permission to use material from the archive of the Eugenics Society. Everybody at Manchester University Press has shown enormous professionalism, patience and hard work.

I would particularly like to thank for their great help and encouragement: Mariateresa Boffo, Esperanza Brizuela Garcia, Raj Brown, Peter Cain, Nancy Campbell, Jeffrey Diamond, Caroline Elkins, John MacKenzie, Rebekah Polding, Andrew Porter, Ruth Prakasam, Richard Rathbone, Tabassam Shah, Robert Tombs and Simon Winder. Above all, thanks must go to Pasco, Rosa and the rest of my family for all their support.

GENERAL EDITOR'S INTRODUCTION

The development of pseudo-scientific racism and the possession of empire
were, in the British case, inseparably intertwined. Robert Knox, the notorious
Edinburgh anatomist, derived many of his racial ideas from the period when he
was a military doctor in the Cape Colony participating in the frontier wars. Later,
Francis Galton combined his interests in genetics, statistics and anthropology
from the studies he made while travelling extensively in Africa. James Hunt
considered that the possession of an empire should prompt the British to a close
study of race, while Henry Flower was a confirmed proponent of the central need
for anthropological research. Any analysis of the sessions and addresses at various
of the disciplinary sections of the British Association for the Advancement of
Science reveals just how prominent were the concerns of scientists with the
apparent opportunities and perceived responsibilities attendant upon the existence of the British Empire.

The science of genetics and the social variant which Galton called eugenics
started out largely without any racial connotations, though class was always
central to its studies. The racial dimension was however soon added, not least
because this pseudo-scientific theory came to be disseminated within imperial
territories and began to develop new theoretical positions, not only in relation to
race, but also in respect of notions of criminality and psychiatric illness. Chloe
Campbell here examines, in perceptive and penetrating detail, the manner in
which a group of doctors began to put together this baleful set of ideas in the
British colony of Kenya. Their work was very much based upon late-Victorian
and Edwardian research, but was developed particularly in the years between the
two world wars in the twentieth century. Kenya was a cockpit of such research
for a number of reasons. It had a white settler population which was interested in
such concepts as a way of supposedly legitimating and regulating their position
in the territory and their relationship with its indigenous peoples. It had prisons
and mental health institutions which doctors felt they could use as laboratories
for their eugenics studies. And its profile was sufficiently high that the doctors
were able to establish close relations with scientific institutions and prominent
figures in the imperial metropole. Similar research was undertaken in South
Africa and elsewhere.

From the standpoint of the twenty-first century, it is alarming that this
research and the theoretical positions to which it gave rise found favour with
prominent British scientists – like Julian Huxley – and allegedly progressive and
liberal institutions. But the full-scale adoption of eugenicist theory by the Nazis
had the effect of bringing it into disrepute. It soon encountered resistance in key
areas of British politics and in the Colonial Office. As a result, the efforts of the
Kenyan doctors to secure funding for a major research centre failed. Perhaps
nothing better represents the social, cultural and political contexts in which
pseudo-science (some might argue, all science) flourishes or dies. Yet we know

that eugenics had an 'after-life' in the post-Second World War era in the United States, Scandinavia and elsewhere. Even then it was closely bound up with the phenomena of migration, class formation and survival, and alleged racial difference. It also, shamefully, resurfaced in the era of Margaret Thatcher, when Keith Joseph found its ideas attractive when speaking controversially about the reproduction of what he viewed as a British social residuum.

Chloe Campbell's work reveals that any attempt to separate racist ideas and pseudo-scientific research and practices from the imperial context is doomed to failure. As with so many scientific activities of the nineteenth and twentieth centuries, the existence of the imperial laboratory was key. In the Kenyan case, the development of the Mau Mau campaign in the 1950s meant that some of the eugenicist ideas lingered on, but by then the settler context, and its attempted minority dominance, was doomed.

John M. MacKenzie

CHAPTER 1

Introduction:
Nellie's dance

In July 1933 an aristocratic farmer-settler, the Honourable Eleanor, or 'Nellie', Grant, went to a ball held for the navy by Kenya's Acting Governor, Sir Henry Moore, at Government House, Nairobi. She had a 'lovely sit-out' with the Reverend Wright, Dean of Nairobi, and their conversation turned to some of the big issues of the day, religion and eugenics:

> I gathered he thought the Church, tho' fatuous, needn't necessarily do much harm if you took it the right way. He is terrifically eugenic minded. We stayed v. late, & the daylight hurt a good deal the next mg.[1]

Nellie Grant attended the dance with a party who were all members of Kenya's eugenics society, the Kenya Society for the Study of Race Improvement (KSSRI), whose inaugural meeting had been held earlier that evening in one of Nairobi's most popular meeting places for Europeans, the New Stanley Hotel.[2] The progression over the course of an evening from eugenic committee to grand social gathering reflects how comfortably the eugenics movement fitted with the attitudes of the colony's social and administrative elite. This book explores eugenic thinking in British colonial culture; in particular, it seeks to examine how British eugenic thinking was adapted in a society brutally shaped by racial divisions. It is the story of an intense flirtation with eugenics among a group of Europeans in Kenya in the 1930s which unleashed a set of writings on racial differences in intelligence more extreme than that emanating from any other British colony in the twentieth century.

It should come as no surprise that Britons living in Kenya in the 1930s were influenced by eugenics: eugenic ideas were strikingly pervasive among the British educated middle and upper classes in the first half of the twentieth century, and most of the British inhabitants of Kenya, official and unofficial, came from these classes. What is remarkable about the eugenics movement in Kenya is the strength of its conclusions about race and intelligence, and the ease with which British eugenic principles could be used to construct such extreme scientific racism. So what were the essential principles that underlay British, and later Kenyan, eugenics? Francis Galton (1822–1911), the founding father of eugenics, asserted in an article of 1865 entitled 'Hereditary Talent and Character', and then in his 1869 book, *Hereditary Genius*, what was to become the basis of eugenics: he argued that human beings were unequal, and that these inequalities were hereditary.[3] Galton described eugenics as 'the study of agencies under social control that may improve or impair the racial qualities of future generations either physically or mentally'.[4] Eugenics sought to use modern understandings of science and human reproduction to confront modern social demographic woes; this approach was compatible with a wide range of political and social views and it enabled eugenics to attract a broad constituency of supporters.

Eugenicists wanted to use the powerful concept of heredity to shape and control social change by selective breeding. The eugenics movement in Britain was particularly concerned with hereditary traits associated with social class; central to British eugenics was an understanding of the effects of demographics upon the relative social composition of the nation. It has been convincingly argued that the context of a declining birth rate, in particular a socially differential decline in the birth rate, was central to the development of British eugenics.[5] It was feared that wealthy, professional families were likely to produce fewer children than members of the improvident pauper class who threatened to swamp the British racial stock with a degenerate and undesirable population.

The British Eugenics Education Society, founded in 1907 (renamed the Eugenics Society in 1926), always remained small, elite and slightly obscure; its journal, the *Eugenics Review*, was hardly a best-seller (its circulation was about 1,500 in 1939).[6] However, there is little doubt that eugenic thinking had cultural influences that 'transcended the limited institutional boundaries of a formal organization'.[7] Eugenic thinking permeated many aspects of British cultural life in the first half of the twentieth century; it informed areas of social concern from birth control and the provision of family allowance to mental deficiency and the education of prisoners. Eugenic ideas pulsed through the literary world, parodied by Wodehouse, but also, it has recently been argued, shaping one of the most influential high modernist novels, Virginia Woolf's *Mrs Dalloway*.[8]

Recent work by historians on different eugenic movements across the globe has further emphasised the extent of eugenics' cultural resonance. In his review of recent literature on eugenics, Frank Dikötter gives an indication of the breadth of the movement's cultural significance, pointing out that support for eugenics could be found in countries as varied as 'Brazil, India and Sweden':

> Eugenics was a fundamental aspect of some of the most important cultural and social movements of the twentieth century, intimately linked to ideologies of 'race', nation and sex, inextricably meshed with population control, social hygiene, state hospitals, and the welfare state ... It was part of such widely discussed issues as evolution, degeneration, civilization, and modernity, and touched on a wide variety of emerging fields like maternity, psychiatry, criminology, public health, and sex education.[9]

Recent histories of eugenics in non-English speaking regions and comparative studies of eugenics in different countries have widened our understanding of eugenic thought, breaking away from the narrow, Anglo-American model that tended previously to dominate perceptions of the subject.[10] But imperial eugenic movements have not yet been brought fully into this discussion; I seek to redress this neglect by locating eugenics in the colonial sphere and tracing the transportation and mutation of British eugenic thought as it moved through the imperial conceptual network. Using Kenya colony as a case study, the book will show how the African imperial enterprise generated a novel eugenic project and how eugenic thinking articulated some of the most profound biological anxieties about race and degeneracy in colonial culture. As well as expressing the cultural fears of colonialism, eugenics also expressed the modernity of the colonial project in Africa: the newness of settler society and the perceived rawness of African development presented an ideal opportunity to create a society modelled on eugenic insights. Kenya could be a laboratory for 'scientific colonisation',[11] untainted by the degenerative effects of misguided old-world sentimentality about nurture.

The fuller cultural implications of imperialism have, like eugenics, only recently been explored by historians – part, in fact, of the intellectual project of this 'Studies in Imperialism' series is to develop the cultural history of empire. Central to this book is the idea of empire as a cultural system through which thoughts and practices were exchanged and modified. In the case of the eugenics movement in colonial Kenya, Britain began as the source in this imperial system of exchange of ideas, but a more complex interaction developed between the metropole and colony. The Kenyan eugenicists started to export their ideas back to the British Eugenics Society and scientific establishment when they produced research and theories that were new and distinctive. The Kenyan eugenicists claimed specialist knowledge on the issue of race and intelligence; the appropriation of this niche, unfamiliar to many British eugenicists, for a time gave the Kenyan research a surprising and distorted authority within the British Eugenics Society. This book shows how eugenics and imperialism – two major forces in early twentieth-century cultural history – were intimately connected; how, in fact, eugenics served as a scientific bulwark that fortified the ideology of imperialism.

The dramatic pungency of Kenyan eugenics and the violence with which the African population was problematised originated in the peculiarities of Kenyan settler culture. A frontier mentality and a peculiarly British snobbery and colonial complacency shaped Kenyan settler society; these forces also directed the shape of Kenya's eugenic thought, which was characterised by an aggressive racial prejudice and a bullish sense of its own intellectual importance. The small size of Kenya's settler population (European inhabitants of Kenya numbered 16,812 in 1931[12]) and the relative absence of a professional intelligentsia meant that it was easier for a single, maverick figure to dominate eugenic ideas within the colony without the credentials that in Britain would have been necessary in gaining a similar level of intellectual authority. The importance of the Kenyan eugenicists' ideas also became strangely magnified within the colony because they chimed so powerfully with overwhelming settler preoccupations and anxieties about the African population and its advancement. As we shall see in Chapter 3, the Kenyan eugenicists' theories were transported to Britain in the 1930s and caused considerable controversy and debate on the role of race in science. In retrospect it is surprising that the Kenyan eugenicists managed to be taken as seriously as they were, albeit for a short time, in the metropole. This can partly be explained by the distorting effects of the colonial network: Kenyan settlers tended to be well-connected within the British establishment and the seriousness with which the Kenyan eugenicists' ideas were regarded by the colony's political and social elite meant that their voices, for a time, echoed disproportionately loudly in Britain.

By the 1930s, Kenya's settler community was already notorious in Britain for its 'Happy Valley' set – scandalously decadent and promiscuous aristocrats whose antics in Africa gave the colony a reputation for a raffish drug-addled glamour and eroticism. This chaotic group of *enfants terribles* was in fact merely a conspicuous minority. Yet there was certainly some truth to the perception that Kenya's white settlers tended to be more aristocratic than, for example, those who settled in South Africa or Southern Rhodesia (now Zimbabwe). As David Anderson has put it: 'From the earliest days of white settlement, Kenya established a reputation as a home for the English gentry, with a distinctly military orientation ... While Kenya attracted the officer class, Rhodesia made do with rank and file ... Kenya was a colony for gentlemen, not for artisans or the white working class.'[13] What is more, the racial dynamics of the colony fostered a powerful sense of privileged entitlement in settler society. The cheapness of land and African labour allowed British settlers to re-enact a landed-gentry lifestyle that had dwindled almost to the point of obsolescence in Britain by the early twentieth century. These aspirations were reinforced by the social and political dominance of certain major landowners. The most prominent figures in Kenya's European immigrant community – men like Lord Delamere and Ewart Grogan – demonstrated theatrical individualism and revelled in their maverick frontiersman identity, whilst also enjoying the benefits that came from belonging to the British social elite.[14] These settler-pioneers personified the semi-feudal, pre-industrial social and racial values that shaped Kenya's settler culture and social identity.

The white population of Kenya was not exclusively composed of debauched and drugged aristocratic younger sons and charismatic land magnates acting out a seigneurial fantasy that was no longer sustainable in modern Britain. There were many smaller landowners, typically from a British upper-middle class, ex-military background, who struggled against their own agricultural incompetence and the marginality of the settler economy to make a living as farmers. There was also an increasing urban and professional population. In fact, only 28 per cent of the European population were occupied in agriculture in 1931: 24 per cent were in commerce; 20 per cent in government; 11 per cent in industry; and 13 per cent were missionaries or professionals.[15] By 1931, 50 per cent of the European population lived in Nairobi and Mombasa.[16] Despite this, settler society idealised an imagined colonial rural simplicity, dismissing the intellectual pretensions of modern, urban society in Britain. This distrust of urbanisation and modern urban social relations was a powerful trait among settlers. As a counterpoint, the supposed pre-industrial simplicity of the African population was often eulogised by settlers; perhaps most famously seen in Blixen's *Out of Africa* (first published in 1937).[17]

Settler culture also idealised the strictly maintained but benignly pater-nalistic class relations that were imagined in this pastoral fantasy, and which enabled them to view the African population as a childlike yeomanry who required the authoritative leadership of Europeans. The reality behind these Kenyan race relations was far more sinister. From early on in Kenya's colonial history, the settlers earned the colony a repu-tation for brutality and racial violence that non-settler territories avoided. As David Anderson has put it:

> By the early 1920s, the deaths of several African servants from beatings at the hands of their European masters earned Kenya's white settlers an unen-viable reputation for brutality ... Physical violence was an integral and char-acteristic part of European domination in Kenya from the beginnings of colonial rule, and by the 1920s it was largely engrained as part of Kenya's peculiar pattern of 'race relations'. Happy Valley ... as some liked to call the White Highlands, was always a violent place if you were an African.[18]

The eugenicists in Kenya were mostly urban professionals and government officers eager to form what was considered an 'intellectual' movement. Only a small proportion of members of the KSSRI members were farmers: the majority worked in colonial administration or were professionals based in Nairobi, or were the wives of men employed in these areas. However, the problematic that Kenyan eugenics constructed was still to a large extent determined by the settlers' anti-intellectual frontier mentality, and Kenyan eugenics colluded with these social attitudes by arguing that the brain of the 'Kenya native' was different and inferior, and warning of the dangers of the educated, Europeanised Africans, who were considered to be corrupted by the trappings of civilisation. The vehemence with which the African mind was pathologised by Kenyan eugenics mirrors the violence inherent in the physical reality of Kenyan race relations. Yet this eugenic thinking was not simply the preserve of the most vociferously racially hostile members of Kenyan settler society. At the heart of Kenyan eugenics lies the apparent paradox that it was supported by individuals who were considered progressive, and by some officials who were viewed as suspiciously 'pro-native' by local settlers. The eugenic research on race and intelligence was supported by many of those concerned with native welfare, the 'pro-native' sector of the European community, because it was presented as consonant with the desire for a more informed understanding of African life and the scientifically informed pursuit of native development. This, and an underlying shared consensus about the reality of African backwardness as a colonial problem, made the intellectual project possible. Eugenicists promised to use reason to take the poison out of the debate on race; the promise was ultimately disingenuous, but for a time it was this appeal that gave the movement its unusually broad base of support.

Eugenic theories on race and intelligence were widely supported by the medical profession in Kenya, as well as powerful members of the official and non-official European settler population, although the missionary organisations, which remained silent on the issue, constituted a conspicuous exception to this. In Britain, the Eugenics Society tended to attract people who were experts in related, specialist fields, such as genetics.[19] The main experts on eugenics in Kenya were medical doctors, who either worked in general practice as colonial medical officers or whose areas of specialisation were unrelated to genetic science, like psychiatry (Dr Gordon) and dermatology (Dr Sequeira). Doctors Gordon and Sequeira were not experts in genetics or evolutionary biology, but were of high professional repute in Kenya. Their audience, while consisting of doctors and those who formed the educated, 'literary' section of the settler community, were not in a position to make alternative assertions about the processes of human heredity. Their understanding of eugenics was largely enforced by the knowledge and interests of Dr Gordon in particular. The strong consensus on the question of race and racial difference meant that the agenda of Kenyan eugenics was surprisingly (although not entirely) uncontroversial within the colony. The central task that Kenyan eugenicists set themselves was thus to explore the mental capacity of the East African, seeking explanations for African 'backwardness' in innate, inherited differences in brain structure:

> Quantitatively the inferiority amounts to 14.8 per cent: qualitatively the cells of the new brain compared with those of the average normal European, show defect and deficiency ... Capable enough of the mere acquisition and retention of information such a people may be; it is another matter to attempt the general exercise of its possibly rudimentary, or even absent, capacity for the higher processes of mind as if this capacity was already fully in being.[20]

The combination of Gordon's leadership and the almost complete unanimity of belief in racial differences in mentality (there was some dispute about the extent of this difference, but there was little doubt in its existence at some level) gave Kenyan eugenics its distinctive flavour.

The story of the eugenics movement in Kenya can add to our growing understanding of eugenics as a peculiarly flexible and adaptable set of ideas. It can also tell us something about British eugenics. The intellectual provenance, the complex motivations, and the interaction with British eugenics indicate that, with all its racial extremity, Kenyan eugenics was not simply a bizarre anomaly, as it was later dismissed in the metropole, but a logical extension of British eugenics in a colonial context. The basis for much of the eugenic research in Kenya was the controversial question of mental deficiency, or 'amentia', in the African population. Although British eugenics had not previously shown great interest in race, the incidence, causes and manifestations of amentia were major preoccupations. The Kenyan research, by introducing a racial dimension to amentia, for a time promised to provide a 'missing link' in the evolutionary explanation of mental deficiency. Race and class were easily conflated in the interaction between Kenyan and British eugenics; mental deficiency became the scientific connection that intellectually bound attitudes to the two social problem groups troubling the respective movements: the Kenyan native population and the British social residuum. The Kenyan movement held a mirror up to British eugenics, and the reaction of British eugenicists to this image reveals an intriguing chapter in the history of British eugenics and scientific racism.

Historians of British eugenics have been rightly cautious about the relevance of race to their subject. As mentioned above, race was not a primary concern for the Eugenics Society in Britain; fear about the effects of degeneration emanating from the white social problem group was the main preoccupation in a country that had not experienced substantial non-white immigration. What is more, the hardening and increasing ubiquity of racism in western European culture from the mid-nineteenth century onwards means that it would be distorting to isolate eugenicists as being particularly racist. However, historians of scientific racism have found there to be a definite relationship between eugenics and scientific racial thought in the first half of the twentieth century, and have often seen the context and influences of eugenics as fundamental to their subject.[21] One of the objectives of this book is to create a framework which allows for a reintegration of the apparently conflicting historical approaches of historians working on eugenics and those working on scientific racism. It is argued that in order to understand better the British eugenics movement, we need to recognise how central eugenics was to colonial racial theories, and that in Britain itself, eugenics was deeply connected to the ideology of empire.

The use of the word racism in historical analysis can be problematic. Banton has eschewed the use of 'racism' as being misleading and potentially anachronistic, arguing that its pejorative slant obstructs our view of historical meanings and understandings of race. Banton argues that it is a mistake to isolate any individuals, eugenicist or not, as racist because they were responding to the evidence produced by scientific knowledge at that time.[22] In this study, racism will be understood as the interpretation of differences between distinct human groups as innate and biological, involving a qualitative hierarchy. Scientific racism is a useful term to describe the scientific, biomedical pursuit of theories supporting and explaining this conception of human variance; 'scientific racism' has been used by both Barkan and Dubow in their valuable studies of the role of race in science.[23] The phrase captures the reality that although such biologically based racial thought is now recognised by most as a profoundly mistaken dead end in intellectual history, when placed within its own historical context it was often not considered to be fraudulent or pseudo-scientific. In fact it was considered a valid subject for respected scientists attempting to ascertain biological truths through accepted methods.

This book scrutinises colonial racism through the prism of eugenics and biological science. Kenyan eugenicists sought to use metropolitan eugenics and international racial science to fortify the edifice of white unofficial settler supremacy while at the same time eliciting the support of many of Kenya's most progressive, 'pro-native' officials. The recruitment of these two camps was achieved through the shared belief in the objectivity and progressiveness of science and an underlying consensus about African backwardness. It was made possible by the ambivalence of eugenics itself; eugenic thought was identified with modernity, rationalism and progress, while also reacting against the dangers perceived to be physically inherent in modern social change. It was by constructing a project that sought to use modern scientific knowledge to understand social and demographic problems that eugenics could accommodate individuals pursuing progressive rationalism, and those motivated by a fundamental racial hostility. By analysing the formation and meaning of racial thought, this study sheds light on the colonial *mentalité* and the complex ideological layers and affinities – metropolitan and colonial, settler and official, progressive and conservative – that rather uneasily merged to form a science that could defend the racial system upon which the Kenyan colonial state rested. It reveals how eugenics was intellectual ballast for the ideology of British imperialism.

Notes

1 Grant to Elspeth Huxley, 12 July 1933, RH, MSS.Afr.s.2154/1/1.
2 'Study of Race Improvement', *EAS*, 8 July 1933, p. 46.
3 F. Galton, *Hereditary Genius* (London, 1869). 'Hereditary Talent and Character' is cited by G. Jones, 'The Theoretical Foundations of Eugenics' in R. Peel (ed.), *Essays in the History of Eugenics* (London, 1998), pp. 1–19.
4 F. Galton, *Essays in Eugenics* (London, 1909), p. 35.
5 R. Soloway, *Demography and Degeneration: Eugenics and the Declining Birthrate in Twentieth-Century Britain* (Chapel Hill, 1990).
6 Blacker to Granville Edge, 23 January 1939, CMAC, SA/EUG/C.130.
7 Soloway, *Demography and Degeneration*, p. xvii.
8 D. J. Childs, *Modernism and Eugenics: Woolf, Eliot, Yeats and the Culture of Degeneration* (Cambridge, 2001).
9 F. Dikötter, 'Race Culture: Recent Perspectives on the History of Eugenics', *American Historical Review*, 103/2 (1998), pp. 467–78, 467.
10 For examples, see M. B. Adams (ed.), *The Wellborn Science: Eugenics in Germany, France, Brazil, and Russia* (Oxford, 1990); N. Stepan, *'The Hour of Eugenics': Race, Gender, and Nation in Latin America* (Ithaca, 1991); R. Cleminson, 'Eugenics by Name or Nature? The Spanish Anarchist Sex Reform of the 1930s', *History of European Ideas*, 18/5 (1994), pp. 729–40.
11 H. L. Gordon, 'Amentia in the East African', *Eugenics Review*, 25/4 (1934), p. 235.
12 D. Kennedy, *Islands of White: Settler Society and Culture in Kenya and Southern Rhodesia, 1890–1939* (Durham, 1987), p. 197.
13 D. M. Anderson, *Histories of the Hanged: Britain's Dirty War in Kenya and the End of Empire* (London, 2005), pp. 79–80.
14 See E. Huxley, *White Man's Country: Lord Delamere and the Making of Kenya*, Volumes I and II (London, 1935), and E. Paice, *Lost Lion of Empire: The Life of 'Cape-to-Cairo' Grogan* (London, 2001).
15 Kennedy, *Islands of White*, p. 198.
16 M. G. Redley, 'The Politics of a Predicament': The White Community in Kenya 1918–1932', PhD thesis, Cambridge, 1976, p. 6.
17 K. Blixen, *Out of Africa* (London, 1954).
18 Anderson, *Histories of the Hanged*, p. 78.
19 See G. R. Searle, 'Eugenics and Class' in C. Webster (ed.), *Biology, Medicine and Society, 1840–1940* (Cambridge, 1981), pp. 217–42, and P. Mazumdar, *Eugenics, Human Genetics and Human Failings: The Eugenics Society, Its Sources and Its Critics in Britain* (London, 1992).

20 H. L. Gordon, 'The Intentional Improvement of Backward Tribes', *EAMJ*, 11/5 (1934), p. 235.
21 On the role of eugenics in scientific racism, see E. Barkan, *The Retreat of Scientific Racism: Changing Concepts of Race in Britain and the United States between the World Wars* (Cambridge, 1992); S. Dubow, *Scientific Racism in Modern South Africa* (Cambridge, 1995); and N. Stepan, *The Idea of Race in Science: Great Britain, 1800–1960* (London, 1982).
22 M. Banton, 'Galton's Conception of Race in Historical Perspective' in M. Keynes (ed.), *Sir Francis Galton, FRS. The Legacy of His Ideas* (London, 1993), pp. 170–9, and *Racial Theories* (Cambridge, 1987).
23 Barkan, *The Retreat of Scientific Racism*, and Dubow, *Scientific Racism in Modern South Africa*.

CHAPTER 2

British eugenics, empire and race

The cultural pervasiveness of eugenics, described in the previous chapter, does not always make its influences easy to define: the eugenic programme was largely theoretical, and the intellectual project was always work-in-progress. As a set of ideas, eugenics was profoundly malleable, marked by a deep ambivalence towards concepts of progress and modernity and consequently able to appeal to very different political and social visions. British eugenics was the intellectual mother ship for the Kenyan movement; the priorities, the language and the racial theories adopted or countenanced by British eugenics form an essential background to understanding what made the Kenyan eugenic discourse distinctive and at the same time palatable to British eugenicists. This chapter considers the history of British eugenics, its internal debates and evolving position, and examines what aspects of British eugenic and racial thought influenced Kenyan eugenicists in the formation of their agenda for 'scientific colonization'.[1] The wider imperial implications of British eugenic thought will then be considered, allowing a further examination of the intellectual connections and fissures between colonial and metropolitan eugenic thought.

The British eugenics movement

Eugenic thinking emerged in the last decades of the nineteenth century; the term eugenics was first used by Francis Galton in 1883 in his book *Inquiries into Human Faculties*, but the organisation, the Eugenics Education Society, was not formed until 1907 (it was later renamed the Eugenics Society). Galton, with typical tetchiness, was reluctant to become involved in the Society, which was inspired by his ideas, although he did accept membership in 1908 and was elected honorary president. Membership numbers never got beyond 1,700, but branches sprang up in Birmingham, Cambridge, Manchester, Southampton, Liverpool and

Glasgow.[2] The Society made up for its smallness with the social and intellectual clout of its membership:

> Local British and American groups listed leading townspeople among their members, and the national councils included distinguished scientists and social scientists, prominent lawyers, clerics, physicians, schoolmasters, intellectuals, and – in Britain – several knights of the realm ... meetings of a eugenics society at Cambridge University before the war drew hundreds of people, including high college officials, Nobel Laureate scientists, powerful senior professors, and the young Maynard Keynes ... The president (from 1911 to 1928) of the British society bore a name to conjure with in matters of descent – he was Major Leonard Darwin, a son of Charles Darwin.[3]

There has been intense debate among historians about who and what eugenics politically and socially represented, but there is broad agreement that it was predominantly an educated middle-class movement. Donald MacKenzie has tended to see the eugenics movement as defending professional middle-class interests and closely related it to Fabianism.[4] A similar argument was put forward by Freeden, who strongly identified eugenics with middle-class progressives, in particular the Fabians.[5] Jones has contested the idea that the Eugenics Society was dominated by the professional middle class by arguing that the hostility of British eugenics to the urban poor reflected a capitalist agenda and by pointing to the large number of businessmen in the London branch. Furthermore, the presence of branches of the Society in manufacturing cities such as Birmingham, Liverpool and Manchester has been neglected by historians, according to Jones, leading to a further underestimation of the links between business interests and eugenics.[6] Searle, on the other hand, has argued that eugenics should be seen as both on the right of the political spectrum and as representing a very narrow band of interests, in particular geneticists and others whose professional interests, in both senses, complied with the furtherance of eugenic thought.[7] Searle has also maintained that in terms of the national impact of the Eugenics Society, professional men were represented far more than businessmen, and more importantly, that the Society took very little interest in business concerns, pointing out, for example, that it was rare for a businessman to address the Society or for business perspectives to be represented in the *Eugenics Review*.

As Mazumdar has argued, the Eugenics Society certainly represented a narrow social group – middle-class biological or social scientists, and a proportion of activists who approved of the organisation's social goals – that had a history reaching back to the nineteenth century and the efforts of 'a reforming middle class to improve the lives of their inferiors in the classes below them'.[8] This background of Victorian social reform, Mazumdar suggests, gave British eugenics its particular character. The desire to combat the causes of the social problem group, the urban pauper class, by confronting supposed underlying biological causes was the major objective of British eugenics. Victorian meliorism melded with the theories of heredity put forward by Darwin and Galton and the Malthusian notion of the excessively prolific poor to form the eugenics movement.[9] As will be shown later, these three elements (social reform, heredity and Malthusianism) varied in their success in surviving transportation to colonial Kenya.

Searle has argued that overall, the influence of the Eugenics Society lay less in policy-making, whatever social group it was representing, and more in its role as a 'specialised pressure group which had the courage to speak out on controversial matters from which democratically accountable politicians prudently kept clear'. Eugenics was also influential in its contribution to the advancement of disciplines like demography and genetics.[10] An example given by Searle of the success of the Eugenics Society in acting as a pressure group prodding cautious politicians was the 1913 Mental Deficiency Act. The Act came about following the Royal Commission on the Care and Control of the Feeble-Minded, which was appointed in 1904 and reported in 1908.[11] The Commission stated that approximately 0.46 per cent of the population could be labelled as mentally defective. The passing of legislation on the matter of feeble-mindedness in 1913 was largely thanks to a campaign by the Eugenics Education Society and the predominantly eugenic-minded National Association for the Care and Control of the Feeble-Minded, with the help of publicity in *The Times*. The legislation encompassed many varieties of so-called mental deficiency, ranging from cretinism to an inability to benefit from education. The law established a central Board of Control which had power to compulsorily institutionalise and segregate, to limit the propagation of the feeble-minded. The law was regarded as a triumph for the eugenics movement, and was the only piece of legislation passed in Britain that was clearly influenced by eugenics. The eugenic implications of the Act were heightened by its remarkable definition of mental defect, which included 'not only paupers and habitual drunkards but women on poor relief at the time of giving birth to, or being found pregnant with, an illegitimate child'.[12]

At the heart of British eugenic thought, there were two methods of using eugenics to change the social composition of the nation. These were positive eugenics, which involved the encouragement of marriage and reproduction between those people considered to have desirable qualities, and negative eugenics, which meant preventing those with unwanted hereditary failings from being allowed to pass them on to future generations. Various proposals were put forward to implement negative eugenics including the sterilisation of the unfit (either voluntary or compulsory), segregation and social pressure through education.[13] Mental deficiency, crime and insanity were problems of particular concern as supposed symptoms of hereditary unfitness in Britain. Negative eugenics generally received greater emphasis than positive eugenics, although in the 1930s there was something of a shift as the Eugenics Society developed a greater interest in issues such as child-rearing and the family allowance as ways of encouraging good eugenic practice.

The Kenyan eugenics movement emerged from and formed links with the British Society in the 1930s. In this period British eugenics was facing novel challenges and underwent significant changes as generally liberal and left-wing biologists and geneticists began to criticise the ideological basis of eugenics and the failure of eugenicists to take into account environmental factors in social problems. Previously, some eugenicists *had* warned of the effects of environmental factors, often described as 'race-poisons', which could include urban life, alcoholism and syphilis. However, these arguments were often somewhat muddled as clear distinctions between environment and heredity were missing. The debate on nature and nurture became more polarised during the 1930s as they became more rigorously defined biological concepts. At the same time, the argument became more politicised, as environmentalism tended to become more explicitly linked with the left.

In the 1930s the agenda of the Eugenics Society shifted, partly as a result of the assault on eugenics from eminent figures within the academic bastions which had previously cultivated eugenics. The change has been defined by Kevles as being from 'mainline' to 'reform' eugenics and he has described the reformers in the following terms:

> Some – like Ronald A. Fisher, Karl Pearson's successor as the Galton Eugenics Professor and director of the Galton Laboratory for National Eugenics at University College, London – were antiracist conservatives; others were social radicals in the tradition of George Bernard Shaw and Havelock Ellis. The prominent biologists among them ranged from the moderate left to the Marxist left – from Julian Huxley and Herbert S. Jennings to Lancelot Hogben, J. B. S. Haldane, and Muller. Whether right or left, they were united in recognition that advances in anthropology, psychology, and genetics had utterly destroyed the 'scientific' underpinnings of mainline doctrine, and that any new eugenics had to be consistent with what was known about the laws of heredity.[14]

The chief architect of reform eugenics was Carlos Blacker, who was the General Secretary of the Eugenics Society from 1931 to 1952. A decorated war hero who had 'turned to psychiatry as a result of his experiences in the trenches', Blacker was 'well-connected in medical and scientific circles when he became active in the Society in the 1930s'.[15] The transition from mainline to reform eugenics involved an attempt to lessen the eugenics movement's association with reactionary and extremist interpretations of social problems. Reform eugenics was more cautious when it came to making the kind of crass generalisations about social class associated with some mainline eugenicists.

The interests of the Eugenics Society accordingly changed considerably during the 1930s. Under mainline eugenics, the movement had largely avoided becoming involved in issues such as birth control and child and maternal welfare.[16] In 1926 the Society, after much indecision, had formally adopted the promotion of birth control, primarily aimed at the poor. Formerly there had been anxiety about the issue because it was felt that contraception would only be adopted by the prudent, intelligent, responsible families, the very people who should be encouraged to reproduce. There was also squeamishness about the subject that took some time to be overcome by the more conservative 'mainliners'.

The redirection of the Eugenics Society's interests was indicated by the dropping of 'Education' from the Society's name in 1926, marking a new emphasis on research. Blacker was to turn the Society:

> from a confused, unfocused amateur propaganda organisation dabbling uncertainly in dubious science, into a quasi-professional foundation committed to family planning and the serious study of social and biological population problems.[17]

The Eugenics Society thus became increasingly interested in research questions that focused on social problems from a biological perspective. The bequest of £70,000 to the Society in 1930 from an eccentric Australian sheep farmer, Henry Twitchin, enabled it to fund the Birth Control Investigation Committee, the British Social Hygiene Council, the Marriage Guidance Council, the Joint Committee on Voluntary Sterilization, the Society for the Promotion of Birth Control Clinics and the National Birth Control Association. The Population Investigation Committee was also housed and funded by the Society. At Blacker's urging, grants were given to the Political and Economic Planning group, which was established in 1931 to plan for the recovery of the economy.[18] Through its newly found wealth and Blacker's redirection, the Eugenics Society thus extended its influence into wider areas of social research.

Searle has described a 'resurgence in the fortunes of the British eugenics movement in the 1930s', which continued despite the growing attacks on eugenics and the effects of the Depression.[19] Membership of the British Eugenics Society went up to 768 in 1932–33 and, as Searle argues, it was also attracting prestigious members, which was important to an organisation that aimed at acquiring influence through the status of its supporters rather than mass conversion. The Society aspired to initiate or modify legislation through 'permeating the media, academic institutions and the medical profession'.[20] The highly esteemed members the Society managed to attract at this time included Lord Horder, Lord Dawson of Penn and Julian Huxley.[21] As part of their activities relating to the eugenics movement, these three individuals were all involved in the metropolitan campaign backing the Kenyan eugenic research programme into African intelligence, although with differing degrees of commitment and conviction.

The economic recession of the 1930s, rather than discrediting the Eugenics Society because of the obvious external economic factors that created the sudden rise in poverty and unemployment, temporarily revitalised eugenics. Searle has attributed this resurgence to the fact that eugenic thought bypassed the need for analysis of the wider causes of economic collapse by blaming problems on the degeneration of certain sectors of the population. What is more, the heightened anxiety felt by the middle class about a dissatisfied and enlarged poor, unemployed population led to more antagonistic interpretations of their position. In his 1932 Galton lecture, entitled 'The Social Problem Group as Illustrated by a Series of East London Pedigrees', Lidbetter argued the social problem group in East London was racially distinct:

> The pedigrees reveal that there is in existence a definite race of sub-normal people, closely related by marriage or parenthood, not to any extent recruited from the normal population, nor sensibly diminished by the agencies for social or individual improvement.[22]

Lidbetter's study was based on the collection of family histories, and his pedigree method was criticised within the eugenics movement: the statistician Ronald Fisher attacked the genealogical method for its failure to introduce controls or to factor in the role of environment in social failure.[23] Lidbetter's lecture, its content and the response it elicited, demonstrated how the eugenics movement was being contested internally, and how, although 'reform' eugenics was beginning to change the emphasis and tone of eugenics, in the early 1930s these developments were still in their infancy. Although reform eugenics attempted to see social value in all classes, in some respects the concept of it was, as Kevles has pointed out, self-deluding; notions like 'anti-social character' were still essentially shaped by class prejudice.[24] An indication of the limitations of the Society's reform was its failure, despite Blacker's efforts, to gain serious support in the Labour movement; the 1934 Trades Unions Conference carried an anti-sterilisation motion, in which it was maintained that there would be a class bias in a sterilisation programme.[25]

Despite their milder rhetoric, British eugenicists, even reformists, still accepted the concept of innate biological differences between social groups as useful in interpreting social problems. The Kenyan eugenicists developed connections with and received warm support from the British Eugenics Society in the 1930s; 'reform' eugenics of that time had not yet taken a clear stand against scientific racial thought. One factor in the support of the British eugenicists for the Kenyan research lay in the fact that although dissatisfaction with eugenics was increasingly being expressed, both 'eugenists and their critics were agreed that more research was needed, and both were roughly agreed on the questions that needed to be answered. There was even broad consensus that, eventually, the knowledge gained might be used to upgrade the human race.'[26] The idea that more research into human heredity and matters of social concern should be undertaken was particularly likely to succeed when it came to research in Africa, where it was felt that so little was known about Africans, their social organisations and conditions.

Mental deficiency, or amentia, was a central concern of the Kenyan eugenicists, and this was an issue on which the British Eugenics Society also heavily focused. It is necessary, therefore, to give some background into the development of thinking on mental deficiency at this time. Mental deficiency was the subject of four reports in Britain between 1929 and 1938.[27] The report of the Mental Deficiency Committee (the *Wood Report*) of 1929, on which the Eugenics Society was well represented, sought to discover the incidence of mental deficiency in Britain.[28] Mental deficiency had traditionally been divided into two forms: primary, which was the result of flawed heredity; and secondary, which was the result of environmental factors. Through a survey of six areas, the *Wood Report* found that 8.57 per thousand of the population were 'feeble-minded'. Primary (innate) amentia was more common than secondary, and was found to exist in family clusters, often within the social residuum. The report concluded that the lowest 10 per cent of the population, the social problem group, should be segregated (rather than sterilised) because of its association with feeble-mindedness and concomitant social defects such as alcoholism, criminality, insanity and pauperism.

The second report, the British Medical Association's report of 1932, came out strongly against sterilisation as a means of controlling numbers of the feeble-minded.[29] Although the BMA's Committee on Mental Deficiency, which was responsible for the 1932 report, contained five eugenicists among its twenty-one members, it also included some stout anti-eugenicists. The report placed heavy emphasis on the environmental causes of feeble-mindedness and stated that little was still known about hereditary factors. The third report, the *Brock Report* of 1934, supported the introduction of voluntary sterilisation; this report was produced by the Departmental Committee on Sterilisation, which was chaired by Sir Lawrence Brock, who was also Chairman of the Board of Control.[30]

Finally, the Colchester Survey provided the strongest rebuke to the simple hereditary approach to amentia that sought to use sterilisation and segregation as solutions.[31] By the end of this survey, the 'mainline' approach had been made virtually untenable. By the late 1920s in Britain, some experts on mental deficiency had begun to question the simplicity of the standard classifications of deficiency. One of the people interested in this line of thought was Ruth Darwin, Charles Darwin's granddaughter and a principal of the Darwin Trust. The Trust aimed to assist and encourage research into mental defect and agreed with the Medical Research Council to help fund substantial research into the causes of mental deficiency, in particular to discover whether its causes were primary or secondary.[32]

The doctor Lionel Penrose was chosen to conduct this investigation at the Royal Eastern Counties Institution in Colchester, a major hospital for the mentally deficient, containing over a thousand patients. The research was to establish Penrose's career as one of the leading human geneticists of his time. Penrose was a critic of mainline eugenics and his work on the Colchester Survey did much to undermine the simplistic approach to mental deficiency that had characterised the traditional attitude of the Eugenics Society. His 1933 monograph, *Mental Defect*, criticised the use of the pedigree and the social bias of Lidbetter's research.[33] Penrose emphasised the complexity of causes of mental deficiency and the inadequacy and misguidedness of any simple, universal solution that promised either to prevent or to cure amentia.[34] Mazumdar describes Penrose's Colchester Survey as 'the last grand manifestation of the eugenist problematic. It also marks the beginning of the end.'[35] Through painstaking research into the patients' backgrounds and clinical histories, and by acquainting himself with the individual patients themselves, Penrose accumulated large amounts of evidence. He concluded that the majority of the Colchester cases did not fit the conventional primary/secondary classification, with its overwhelming bias towards primary amentia; instead he found that mental deficiency was generally a combination of pathology, environment and genes.[36]

The Colchester Survey challenged the authority of eugenicist experts on mental deficiency, such as R. J. A. Berry and A. F. Tredgold, who had, as leaders of their field, promoted the mainline eugenics perception of mental deficiency as being caused by 'rotten heredity' – a pedigree of failure that was irrevocably passed through generations.[37] Berry and Tredgold were important influences on the work of Dr Gordon in Kenya; they were also in contact with the Kenyan eugenicists in the 1930s and supported their research. In the early years of the interaction between

British and Kenyan eugenics, the reform movement's enthusiasm for research worked in favour of the Kenyan eugenicists. Ultimately, however, the desire of the British Eugenics Society to distance itself from Nazi racism, the growing momentum of the attack on scientific racial thought and the new critique of traditional approaches to amentia would isolate the Kenyan eugenicists, particularly when the extremism of some of its key proponents was revealed as the rhetoric of the debate on race and intelligence intensified.

Race and empire

In this section, the relationship between British eugenics, race and colonial imperialism will be examined as an introduction to an explanation of how eugenics developed in Kenya and why the Kenyan eugenics research programme attracted the British Eugenics Society. There were two principal ways in which issues arising from empire interested the British eugenics movement. The first was congruent with conventional eugenic fears about the degeneration and decline of the population.[38] It was the idea that being a powerful, imperial nation led to additional pressures on the British population to maintain its racial standard; at the same time, the wealth and comfort of 'civilisation' conflicted with the maintenance of rigorous vigilance against the unfit.[39] It was feared that racial degeneration would lead to a loss of this imperial status and dominance. The other imperial issue for eugenicists lay in the study of indigenous peoples of the colonies and the colonists' relations with them and their environment. The question of miscegenation provided a link between these two elements through the fear that race mixture would lead to the degeneration of the British population.[40] Hence, empire had an ambiguous effect on British eugenicists. It created, as Banton has pointed out, a sense of both supremacy and anxiety.[41]

Eugenics emerged at a time when racist thinking was becoming increasingly dominant: the language of race abounded in eugenic discourse. By the end of the nineteenth century, racism was 'firmly established in popular opinion and science'.[42] Rich has examined the role of imperialism in the formation of racial theories and their influence on British politics, arguing that the British 'liberal imperial model' contributed to the avoidance in mainstream British opinion of the systematic racial theorising that characterised much continental European thought in the inter-war period. However, although the British policy rejected formal segregation, there were 'important racist dimensions' to British imperialism:

when the colonial and paternalist trappings of these models of imperial control were removed and the model transposed in a neat form back to the imperial metropole, then its more markedly racist aspects could be observed close at hand.[43]

Rich went on to argue that this colonial racism reinforced contemporary anthropological perceptions of race difference and strengthened support for restrictions on the emigration of different races out of their countries of origin and a concomitant growth in the desire to prevent miscegenation. In John MacKenzie's work on propaganda and empire, he has further described some of the linkages between imperialism and racial ideas: 'Concepts of race were closely related in popular literature to the imperative of conflict between cultures, and the evidence of superiority it provided. Colonial heroes became the prime exemplars of a master people'.[44] MacKenzie has argued that in the 1920s and 1930s 'an implicit imperialism … underlay most propagandist and entertainment output'.[45] The various contributions to MacKenzie's edited book on imperialism and popular culture indicate that imperialism was central to many aspects of British culture in the first half of the twentieth century, such as children's literature and popular entertainment.[46] Issues of racial difference in British culture were largely accessed through empire, and as Semmel has argued, there was considerable slippage between the rhetoric of imperialism, social Darwinism and patriotism in the early twentieth century.[47]

When implanted in imperial territories, the full implications of this culture of imperialism were borne out in the elaboration of racial theories which often deployed the language of eugenics and social Darwinism. To take South Africa as an example, the concept of racial segregation was only a small step away from the sense within the eugenics movement of national and racial competition and the belief in different levels of racial development. In 1933 British eugenicist Aikman promoted segregation:

> Difficult as it undoubtedly is, some form of mass-segregation of races seems to be desirable but by this term, I do not mean complete segregation. The ideal would seem to be that teachers, administrators, judges and doctors should have access to the more backward races …[48]

Aikman's article was published in the *Eugenics Review*, a specialised journal with a small readership. But such ideas were to express a far more powerful and popular sentiment in the colonial sphere. Dubow, for example, has argued that in South Africa the theory of segregation 'was strongly associated with the demise of mid-Victorian liberalism and the ascendancy of racial science, in particular eugenics'.[49] Eugenics in Britain may not have been especially preoccupied with race, but it supplied the basis for a scientific justification of racist thought.

The idea of racial inequality, just like individual inequality, was taken for granted by mainline eugenicists. This was expressed by Karl Pearson, Sir Francis Galton's successor in the process of establishing the science of eugenics:

> There is no natural equality of human races, any more than there is any natural equality of human beings; they are the product of their past evolution moulded by selection and heredity. As far as we can understand it evolution is largely an irreversible process.[50]

Thus racial competition was seen as a vital factor in the process of natural selection, which, according to the eugenicists, was the most powerful and only lasting means of changing or developing human beings. Pearson dismissed the notion that competition between individuals was of primary significance and went on to refer to racial conflict:

> As soon as we interpret the facts of history in the light of progress by inter-racial struggle, and only in a secondary manner by intra-racial competition, those facts become deeply suggestive and significant for our guidance.[51]

Such a view had significant implications for imperialism. Karl Pearson's social Darwinism was a key expression of social-imperialism in Britain.[52] According to Pearson, the inevitability of international and interracial conflict demanded that Britain should reach its maximum national efficiency and minimise the weakening effects of domestic class divisions: a strong, socialist government, with the aid of nationalism and imperialism, was the best way to achieve this.[53]

A common theme in eugenics was to interpret human history as the story of the rise and fall of different races. Thus according to one such history, the Italian renaissance was caused by an influx of superior Northern European stock into Italy.[54] Equally, the rise of empires in the past were attributed to the superiority of the imperial race, and the decline to racial degeneration. This obviously provided an important lesson for Britain's empire. The eugenicist Saleeby wrote extensively on this subject, arguing against notions of racial senility and the Lamarckian theory of the transmission of acquired characteristics as the cause of racial decline. Instead, he attributed it to the complacency of powerful civilisations, which explained why imperial peoples, in his view, tended to degenerate more than enslaved ones. According to Saleeby 'civilizations and Empires' were more likely to succumb to degeneration because the cumulative external trappings of tradition and education disguised the most crucial factor, the quality of the racial stock:

> Natural selection is the sole factor of efficient and permanent progress, but the traditional or acquired progress which we call civilization tends to thwart or abrogate or even invert this process. I thus believe that the

conditions necessary for the *secure* ascent of any race, an ascent secured in its very blood, made stable in its very bone, have not yet been achieved in history: *and I advance this as the reason why history records no enduring Empire*.[55]

Such fears about the transience inherent in imperial power led to ambivalence about the colonised people; their inferiority was repeatedly asserted but there was also a eugenic respect for a culture in which natural selection was thought to work unimpeded by the luxury or morality of civilisation.

Carlos Blacker contributed in 1952 to the debate on British eugenics and racism through his discussion of Galton's attitudes:

> The theories and practices of racism have opened a veritable Pandora's box of national recriminations and political hatreds. From this turmoil has emerged the doctrine of racial egalitarianism which is orthodox to-day. The reaction is understandable, but Galton might have thought that, if construed to mean that all human races have an equal power of sustaining and promoting the process known as Western civilisation, it belied some of the plainest facts of human evolution ... Galton's views on race as expressed in *Hereditary Genius* are the first tentative efforts of a kindly man to apply to humanity the harsh lessons of biology as then understood. His motive was to prevent, not to inflict, suffering.[56]

In his description of Galton, Blacker captures the attitude towards race that characterised 'reform' eugenics: he displayed a dislike of racism and racial hostility, while finding it difficult to accept the idea that all races had equal mental capacities.

Julian Huxley held a similar position in the 1930s. The process of questioning eugenics, and for our purposes here questioning the racism of the eugenics movement, was not a straightforward one. An example of this can be seen in the development of Huxley's approach to race and genetics. Huxley was a member of the Eugenics Society and although he became a critic of its flaws, he never rejected it; in fact he became President of the Society in 1962. As Barkan suggests, Huxley's role in the movement was important:

> Perhaps his greatest influence came from his unofficial role as a member of the inner circle, involving consultations with Carlos Paton Blacker ... who had been his student at Oxford after World War I, and others who could be termed the 'progressives' in the society.[57]

As a liberal, in the 1920s Huxley espoused some extremely racist ideas that were in sharp contrast to his later well-known antiracist stance. In 1924, after returning from travels in the USA, Huxley wrote a series of articles for the *Spectator* in which he asserted that 'the negro mind is as different from the white mind as the negro from the white body'.[58]

In 1929, Huxley went to East Africa as an advisor for the Colonial Office's Advisory Committee on Native Education in Tropical Africa. After this trip he wrote the book *Africa View* in which he began to question the biological validity of race. He had not, however, reached the stage of being egalitarian about race:

> I am prepared to believe that if we ever do devise a really satisfactory method of measuring inborn mental attributes, we shall find the races of Africa slightly below the races of Europe in pure intelligence and probably certain other important qualities. But – and the but is a big one – I am perfectly certain that if this proves to be so, the differences between the racial averages will be small; they will be only an affair of averages, and the great majority of the two populations will overlap as regards their innate intellectual capacities.[59]

Barkan describes a 'duality' in Huxley's ideas about race in the early 1930s: 'At this stage Huxley was non-racist in his descriptions and analyses, but continued to take a racist stand when it came to conjectures.'[60] In 1935 Huxley published, with the anthropologist Alfred C. Haddon, the book *We Europeans: A Survey of 'Racial' Problems*.[61] This text, which became a best-seller in 1936, was a deliberate attempt to reveal the misconceptions of racism and the fallacy of Nazi racial theories. It was largely an attack on German racism and concentrated on racial questions within Europe. It argued that the notion of a genetically pure race was impossible and emphasised the importance of both nature and nurture in shaping human behaviour and characteristics; its argument was that the concept of race made little biological sense. The word 'race' was described as so problematic that it would be better to use the term 'ethnic group'.

Anthropology, particularly physical anthropology, influenced the role of race in science in the inter-war period. Two of the major British protagonists in this area, Arthur Keith and Grafton Elliot Smith, publicly supported the Kenyan research. Both Keith and Elliot Smith were eminent anatomists who became interested in physical anthropology later in their careers. Both were associated with the British eugenics movement, although they disagreed vehemently on many issues, exemplified most famously by the debate over Piltdown Man.[62] Keith believed that human races were ancient and distinct from one another, emphasising the biological essentialism of race. Elliot Smith was an ardent believer in diffusionism, the theory that cultural and technological inventions were made once and then spread across human populations. The underlying focus of diffusionism was on culture rather than race; it implied that human brains were largely the same in different races and that cultural development relied on the circumstances of exposure and environment. Despite contradictory statements on race and biology, Elliot Smith was a significant figure in resisting the use of racial typologies, making him 'one of the more important early anti-racists'. By the 1930s, Elliot Smith and Keith were reaching the end of their careers. Under the leadership of Malinowski, a new generation of social anthropologists was by then developing which 'largely ignored race'.[63] According to Barkan, race was a matter of growing academic confusion for anthropologists in the inter-war period, which resulted in 'widespread skepticism, but concurrently enabled racist theories and attitudes'.[64] In fact, in the case of the Kenyan eugenicists, Malinowksi was openly critical and scornful.[65] But still the question of the physical basis of racial differences was increasingly left to biologists to resolve. The appeal of the Kenyan eugenicists, who proposed to search for clarity on the question of race and intelligence through scientific examination, was therefore not immediately diminished by the growing uncertainty surrounding race and biology in the discipline of anthropology in the 1930s.

The association between eugenics and scientific racial thought must partly be understood historically; studies of eugenics in different countries have made this increasingly clear.[66] Eugenic thought in France, for example, was identified with an environmentalist approach to child welfare and social improvement. In Germany, many eugenicists tried to avoid an espousal of the Nordic racial myth until Hitler came to power in 1933 and *Rassenhygiene* (racial hygiene) finally became co-opted into Nazism.[67] Eugenics in the Soviet Union became linked with social revolution until it was banned in 1931 with the rise of Lysenko's Lamarckian conception of heredity. Brazil provides perhaps the most intriguing example of how flexibly and unpredictably eugenics was transposed into different political cultures. Here, in an inversion of the British conception of the relationship between nature/nurture and racism, which connected environmentalism with antiracism, the Brazilian Mendelians were associated with an antiracist stance, while the environmentalist neo-Lamarckians represented the more overtly racist political position.[68] Adams has powerfully argued that through an international, comparative approach to eugenics, an image of a more complex and divergent scientific phenomenon emerges. Adams has attacked as a myth the idea that eugenics was '"really" about race or class', or political reaction, because in different areas, eugenics became associated with different political and social interests and ambitions.[69]

Although historians of the British eugenics movement have pointed out the dangers of overstating the link with racism, eugenic thinking was extremely fluid and eugenics' epistemology of humanity and its typologies easily fitted racial arguments. Eugenic thinking in Kenya in the inter-war period illustrates the ease of the transposition of British eugenics to racial concerns. The relationship between Kenyan and British eugenicists demonstrates that although racial biological essentialism may have eventually proven incompatible with eugenics and population genetics, the process by which this was learnt and accepted by the movement was a slow and ambiguous one.

The issues brought up by empire for British eugenics changed over time. In the early period, particularly before World War One, when the language and theories of social-imperialism were at their peak, eugenic ambitions focused on grand themes, often historically, like the rise and fall of empires and the necessity of maintaining an imperial race. In the inter-war period, British eugenicists became more utilitarian: there was more speculation on specific problems arising from colonial-imperialism, in particular miscegenation and the effects of migration. In the years immediately after World War One, when it seemed that there would be an increase in migration from Britain to the colonies, there appeared a flurry of articles on related issues in the *Eugenics Review*. Leonard Darwin wrote in October 1919:

> Immigration and emigration certainly stand first for consideration amongst imperial problems, for we all see the importance of keeping up the quality of the human marble which we are constantly exporting ... and of seeing that this export does not reduce the quality of the human stone which will have to be used in future for the building of our own nation.[70]

Emigration to the colonies raised important eugenic concerns. One was that the departure of healthy, adventurous young men would leave a higher proportion of the defective and unfit population at home to reproduce in conditions of reduced competition for sexual selection. Soloway locates this concern in the context of an already declining birth rate and cites the rise in colonial emigration in the decade before World War One from more than 200,000 a year to over 300,000 a year. This worried eugenicists like Saleeby, who was aghast at Kipling's recommendation that young men and women should go to populate colonies and warned that Britain would be in as dire a condition as Ireland if it continued to encourage its best stock to leave.[71] It was also feared that emigration would encourage miscegenation, particularly because of the disproportionate number of men moving to colonies.[72]

The eugenics movement in Britain was different from its American counterpart in its lesser emphasis on the issue of race. British eugenicists tended to focus more on hereditary differences between social classes. There were, however, clear parallels in the attitudes held towards what were considered inferior races or inferior classes:

> The very best environment has no power in itself to raise a bad or unfit race of people, as for example, gipsies or negroes, and it is in the same way

impossible to raise in civilised societies the deepest sunken bottom layer, i.e., habitual criminals, vagabonds, imbeciles (individuals deficient in gifts).[73]

Eugenics is crucial to our understanding of scientific racial thought: British eugenics may not have been shaped by racism, but scientific racial thought in Britain (and in Kenya) was certainly shaped by eugenics. As Stepan has put it:

> For the student of race science, eugenics is important because it linked race with hereditarianism and the new science of genetics … and because the eventual collapse of racial typology was intimately associated with the collapse of eugenics in the 1930s. No history of British racial science would therefore be complete without some account of the eugenics movement.[74]

The case of eugenics in Kenya demonstrates how it served as an intellectual mechanism for justifying and crystallising racial assumptions.

Eugenics and Kenya colony

Mazumdar has identified three main strands of thought from which British eugenics was constructed: heredity, meliorism (social improvement) and Malthusianism. These strands did not all survive in Kenyan eugenics, and this change in emphasis sheds some light on the differing conditions and preoccupations for the colonial movement. Heredity was central to the Kenyan eugenicists' conception of the innate racial differences in intelligence, but there was little discussion about the possibility of harnessing the power of heredity to modify the population; it was simply regarded as the biological mechanism by which deficiency was maintained through the generations. Meliorism was crucial in motivating some members of the KSSRI: some of the most active social reformers in the colony were eugenicists. Malthusianism, on the other hand, did not have a significant role in Kenya eugenics; the different demographic relationship between Kenyan eugenicists and the African population, and European dependence on African labour meant that eugenics in Kenya was forced to reach different conclusions.

In the 1930s there was little concern about the speed of African population growth. Although indications of increase among Kikuyu in areas like Kiambu were being detected, and some research was undertaken on this in the early 1930s for the Kenyan Land Commission,[75] population growth was not yet considered a problem of broad social concern; in fact there was some concern about under-population. There were exceptions to this, such as Dr Paterson, who became Director of Medical Services in 1933 and who was one of the more liberal and progressive members of the Kenyan eugenics movement. Paterson, whose ideas will be discussed more fully in Chapter 3, issued dire warnings about the growing African population and a potential Malthusian crisis. But generally, although the Kenyan eugenicists' discourse on the hereditary failings of the African population had many parallels with the attitudes towards the pauper class found in British eugenics, there were no comparable expressions of concern about a rising tide of the African social residuum. The government statistician estimated the African population in Kenya to be about 3.15 million in 1932;[76] the European population was 16,812 in 1931.[77] Demographic figures for Africans before 1948 are notoriously unreliable; they were based on the crude figures produced by the hut tax collections. The first official census of the African population was taken in 1948 and counted 5.25 million.[78] The substantial increase in population suggested by these figures is exaggerated because of the inaccuracy of the statistics collected before 1948. However, it has been estimated that after a period of stagnation, or possibly decline, it is likely that the population of East Africa started to rise in the mid-1920s; this rise continued at a 'slower, but steady rate' until the end of the 1930s.[79] It was after World War Two that fears about over-population became more widespread: Gordon wrote an article on the subject in 1945, which he failed to have published in the British press.[80]

Urbanisation, however, was one issue related to demographic change that did feature in Kenyan eugenics. Urban life was seen to lead to 'Europeanisation' and detribalisation, which disturbed African social and psychological stability. The worry was not so much about an increase in the African population as its geographical distribution and the cultural effects of this change. Hence Kenyan and British eugenics converged in their imagery of a dispossessed, criminal and unproductive urban under-class:

> Colonial perceptions of the negative consequences of African urbanisation are coloured by a sense of the 'other' – resulting from an ignorance about African life in the towns – that corresponds with the fear of the unknown inhabitants.[81]

A counterpoint to this anxiety about urbanisation was the eulogising of the pre-industrial simplicity of the African population, a sense that the African was mystically in touch with the natural world, manifested for example in Karen Blixen's bizarre tale about Africans being able to commit suicide by will alone.

Although Kenyan eugenics was dominated by the idea of racial heredity, the eugenics programme made no attempt to classify or divide the African population through the attribution of differing hereditary qualities. Kenyan Africans were not classified by the eugenicists in terms of tribe, which was the most important administrative classification for

Africans at the time. The factors that singled out groups for attention tended to be environmental, such as educational experience, literacy and the disquieting acquisition of apparently European social or cultural aspirations. The fact that the idea of limited educable capacity was applied to the entire African population of Kenya made the question of positive or negative eugenics irrelevant within the African population. The problem of mental incapacity was seen by the Kenyan eugenicists as hereditary and immutable, and they attributed it to the dysgenic effects of African conditions, where cultural standards were described as so low that the deficient were able to flourish and abundantly reproduce, making their level of intelligence normative. In the attribution of the problem to heredity causes, Kenyan eugenics followed the British pattern. The Kenyan eugenics movement was more unusual in the absence of a programme for the improvement of the population. Negative eugenics was impractical because the supposed incapacity affected the entire African population; for the same reason, there was no mention of implementing a policy of positive eugenics that would encourage more intelligent Africans to have more children.

It is entirely predictable that the eugenicists in Kenya should have been so preoccupied with race; it is perhaps more surprising that in a country so dominated by racial issues and racist attitudes, theories of scientific racism did not emerge earlier. Dubow wrote in reference to the tardiness of the development of South African theories of scientific racism: 'there was never a "critical mass" of like-minded thinkers with the capacity to create a firm institutional base for the propagation of their theories'.[82] Similarly, the size and interests of Kenya's European population meant that there was no sizeable intelligentsia. Eugenics generally emerged out of urban, professional or intellectual establishments.

The European population in Kenya was not well established by 1931; 50 per cent of the Europeans had been in the country for less than five years.[83] Redley cites the increasing numbers of European women moving to the colony as evidence that the settler population was beginning to stabilise; the ratio was 790 females per thousand males in 1931, compared with 570 in 1911.[84] More European children were being born in the colony as more women moved there, but Kenyan settler culture was still in its infancy, and as a result, the question of the future tenability of white society was a major preoccupation. In *Islands of White*, a comparative study of settler culture in Kenya and Southern Rhodesia, Dane Kennedy makes extensive use of settler papers, letters and diaries to reconstruct the formation of a settler culture forged by anxious and defensively hostile responses to the alien African environment and its indigenous peoples. Kennedy has drawn attention to the marginality and vulnerability felt by Europeans, despite their political and economic supremacy. For example, fears about the climate led to the emergence of practices such as wearing special red vests, felt-lined hats and spinal pads to protect settlers from the effects of solar rays:

> According to this theory, the intense actinic radiation in the tropical climate was capable of paralyzing and even destroying the nerve cells of Europeans, thereby generating various neurasthenic symptoms ranging from depression to 'outbursts of passion'.[85]

Kennedy points out that worries about the effects of the African climate were more pronounced in Kenyan culture than in Southern Rhodesian, suggesting that this can be explained by the fact that Kenya's white population was more recently established and fearful of the African environment than Southern Rhodesia's.[86] Kennedy also explores the need of settler society to draw strict physical boundaries between settler and native, describing how the threat of African male sexual violence towards European women became a powerful vehicle for expressing fear about contact between the two groups and the dangers of miscegenation. The Kenyan eugenics movement needs to be understood in the context of this wider racial anxiety. These fears about the vulnerability of the European body were deeply physical in nature, and eugenics engaged with the issue of racial difference at a similarly biological level. Eugenics confirmed that the African mind was different from the European mind in a profoundly immutable and physical way. Kenyan eugenics provided a reassuring verification of the innate difference between coloniser and the colonised, but also inevitably echoed the sense of being biologically out of place that haunted Kenyan settler culture.

The question of how unified a community the Kenyan settlers were has been the subject of some debate among historians. Kennedy has argued that racial identity and anxieties created solidarity that united different white interest groups.[87] Kennedy took issue with much of the literature produced on settlers in the 1970s that emphasised the complexity and the economic and political divisions of Kenyan and Southern Rhodesian settler society, such as Mosley's study of settler economies, which demonstrated the variations and clashes of settler economic interests.[88] Mosley described the original division as being between the large concessionaires and the small farmers; as the Kenyan economy developed, intra-settler conflicts reformed along sectoral lines 'between producer and producer, or producer and consumer'.[89]

In an important study of the white Kenyan political arena, Redley examined the factious and complex splintering that characterised settler politics, caused by the politicisation of economic interests. As these economic interests became more complex and diverse, settler politics became increasingly inefficient and fragmented. The protection of interests in relation to other Europeans became higher on the immediate political agenda than notions of racial unity.[90] In contrast, Kennedy has argued that the wider shared characteristics of settler society must be understood in order to understand the settlers' experience and social concerns, and that historiographical emphasis on the internal fissures of the European community has led to blindness to the remarkably powerful sense of racial solidarity.[91] The characteristics of settler culture described by Kennedy – the shared racial fears about the effects of the African climate and the violence of 'black peril' – created a semblance of social unity, transcending the class and status divisions that might threaten racial solidarity.

Duder and Youé have since questioned the homogeneity of settler culture described by Kennedy, suggesting that the claims of egalitarianism within settler communities need to be treated more critically.[92] Closely studying the many letters of Arnold Paice, a British settler farmer in Nanyuki, they point out that the much-vaunted eclecticism of the social life of the small and isolated British Nanyuki settler community was in fact rather disingenuous. In reference to Kennedy's citation[93] of Paice's comment that there were few other places where a 'dirty butcher' might be entertained on one night, followed by a 'general with a string of decorations' the next, Duder and Youé argued that this apparent cohesion and lack of snobbery was deceptive, as the provenance of both butcher and general was the English gentry. They suggested that the political and social alienation of the Afrikaners – a substantial minority within Kenya's white population – belied the apparent classless unity of white settler culture. They further argued that settler unity was undermined by the importance of locality in the identification of political interests, a division also described by Redley. The settlers of Nanyuki are depicted by Duder and Youé as having little sympathy for the grander ambitions of self-government held by the settler leaders: 'Nanyuki had a separate political consciousness, albeit white and propertied.'[94] This book examines how eugenic thought sought to overcome these contradictory impulses in Kenya settler society: the need for racial solidarity and the inevitable splintering of social and economic interests as the colony developed. Kenyan eugenics used a biological account of racial differences in mentality and set out biologically based policy prescriptions for colonial rule to accommodate both the imperial case for 'native development' and promote the long-term viability of settler society while managing the emergence of a white underclass.

The Kenyan eugenicists' emphasis on the African population as the main focus for their original research is a reflection of the socio-economic demographics of the colony. Their approach on the issue of poor whites was heavily influenced by South African eugenic thinking. In South

Africa, where there was a much larger and more socially diverse white population (South Africa's white population numbered 1.8 million in 1931, of whom 300,000 were considered to be very poor[95]), eugenicists heavily emphasised the dangers presented by the 'poor white problem', particularly after the 1929 economic depression, when 'poor whiteism' emerged as a major political issue.[96] As Dubow has pointed out, in South Africa: 'Concern over the moral and physical degeneration of urban blacks was primarily motivated by its effect on whites.'[97] A formal South African eugenics movement emerged after World War One and remained prominent until the mid-1930s. Its most active proponents were an eminent husband-and-wife team: Harold Fantham, professor of zoology and comparative anatomy at the University of Witswatersrand (he in fact set up the Department of Zoology at Witswatersrand and was Dean of the Faculty of Science between 1923 and 1926), and Annie Porter, a senior lecturer in Parasitology at Witswatersrand.[98] Fantham and Porter followed the mainline eugenicist tradition, expressing deep concerns about racial degeneration through the differential birth rate and the rising incidence of feeble-mindedness. Additionally, fears about white supremacy being undermined through miscegenation and the growth of a black urban proletariat were major South African eugenic preoccupations.

Fantham and Porter founded the Race Welfare Society (RWS) in Johannesburg in 1930, three years before the formation of Nairobi's KSSRI: the Great Depression provided an impetus for eugenic movements in the colonies as well as in Britain. The RWS seems to have had a similar following to the KSSRI; both memberships largely consisted of educated male professionals, and early public meetings of both societies were enthusiastically attended and reported in the local media.[99] However, the RWS attempted practical eugenic interventions in a way that the KSSRI did not:

> Between 1932 and 1937 the RWS opened five birth control clinics in downtown and working-class neighbourhoods in Johannesburg, three for white and two for black women. Moreover, it initiated an impressive propaganda campaign promoting modern, medicalised birth control at public lectures and symposia which were frequently reported in prominent newspapers. These events resulted in significant popular support for both eugenics and birth control in the early 1930s.[100]

The KSSRI enjoyed a high profile in Kenya's newspapers and similarly popularised eugenics through public lecture and pamphleteering, but it was not associated with birth control to the same extent as the RWS. This difference lay in the fact that 'poor whiteism' was not quite as live a eugenic issue in Kenya as in South Africa. The smaller, less established white settler population of Kenya was in the early 1930s still hoping to achieve independent self-government. Integral to the political debate about settler self-rule was the issue of native rights and native development, and whether the main priority of the Kenyan state should be native advancement under colonial rule emanating from the metropole, or movement towards an autonomous settler-controlled state along the lines of South Africa. The Kenyan eugenicists' argument that the Kenyan native was incapable of development is linked with this debate about the future of the colony. For Kenyan settlers, the perceived threat came from the competing claim of native development and the fear that this might take precedence over the entrenchment of settler society. In this context, it is no coincidence that the Kenyan eugenicists were more interested in discussing the inherent biological limitations of the native population and the futility of native development than in emphasising the degeneracy of settler stock.

Where the RWS instigated practical eugenic measures by encouraging birth control to shape the social and racial balance of the white population, the KSSRI were attempting to undertake and promote scientific research to prove the inferiority of the African brain. The Kenyan eugenic programme was, therefore, peculiarly academic. At a time of major economic retrenchment due to global depression, the more speculative nature of Kenyan eugenics was a serious limitation. No matter how much the Kenyan eugenicists emphasised the policy implications of their so-called discoveries about African mentality, the fact was that investment in this research was never going to be a high priority, particularly when the Kenyan eugenicists looked to the sceptical British government as a source of funding; as we shall see in Chapter 4, the government was well aware of settler interests and alive to the possible political implications of Kenyan eugenics.

This emphasis on research, did, however, mean that the Kenyan eugenics movement coincided with the emergence in the 1930s of a more general imperial interest in monitoring the state of African conditions, as Vaughan has pointed out:

> This was the decade in which social and scientific research on the problems of African colonies really took off. In the Central African colonies there were investigations undertaken on many aspects of African life, which particularly focused on the changes wrought by over thirty years of British rule ... Underlying all of this research was a deeply felt colonial fear, expressed more and more anxiously in the 1930s, that the 'disintegration' of the 'traditional' structures of African societies was endangering social control, that industrialization, education and urbanization contained within them the seeds of disaffection.[101]

The emphasis on the effects of education and development gave the

Kenyan eugenics movement currency and intellectual legitimacy in metropole and colony. The Kenyan eugenicists were aware that research for Hailey's *An African Survey* was taking place in the mid-1930s.[102] It was hoped, vainly, that Hailey's survey might dovetail in its conclusions and provide publicity for the Kenyan research. Other research conducted at this time included nutritional surveys, in particular those undertaken by the Culwicks and Platt in Tanganyika and Nyasaland,[103] which were crucial contributions to the early history of the study of nutrition. These surveys were multidisciplinary; the Nyasaland team included a medical officer, a botanist, an agriculturalist and an anthropologist.[104] Such research was informed by a progressive, reformist agenda that was manifested most notably in *An African Survey*, but also in other smaller-scale colonial studies that appeared in the 1930s, as described by Vaughan. Hetherington's book on British paternalism in Africa identifies trusteeship and faith in Britain's role in modernising and developing Africa as central to the paternalist ideology of British intellectuals interested in Africa in the inter-war period. Debates about trusteeship were complicated by doubts about the African capability for advancement.[105]

The proponents of the Kenyan research into mental capacity saw their eugenic approach as highly congruent with this application of science to public welfare and understanding. An editorial in the *East African Standard*, the colony's main newspaper, hailed the formation of the KSSRI for using the scientific principles of eugenics both to measure and to direct the evolution of change within African society in Kenya:

> The African cannot be left out. If white civilisation is to succeed in its task in this Continent it must provide for the Native a natural place in the plan. In its simplest form, the proposition is capable of one answer; How is it possible to understand the meaning of the venerable African institutions and beliefs that western civilisation is intending to change unless at least some of those who have undertaken the responsibility have some knowledge of psychology, of elementary physiology and the biology of sex reproduction and the influences of environment and heredity? If principles, established by science, have to be applied to the plant and animal worlds intelligently and increasingly if there is to be any improvement in type and productivity, how much more necessary is it to apply knowledge to man himself to prevent the degeneration of the stock through preventible causes? Here is a task to hand in East Africa – a task big enough to occupy the time and thought of generations – and a Society which deserves and requires the support of all people concerned in the welfare of the race. Much knowledge is going to waste in East Africa, Government does not encourage or assist its servants or others engaged in the work of development to place on record the results of their intimate studies of the African race. It is a curious reflection on the Briton in East Africa that there are far more books produced on the animal kingdom than on the human millions

in the territories. Yet although the lion and the elephant are relatively static, mankind is changing almost daily and we have no accumulated date by which to measure the rate and direction of that movement.[106]

The formation of a eugenics movement in the colony was seen as heralding a new era of science and social responsibility which represented an intellectual coming-of-age in settler society. Thus the tradition of social meliorism, described by Mazumdar as a powerful strand in British eugenics, was clearly invoked by the Kenyan eugenicists. British Malthusian arguments on the other hand did not directly inform Kenyan eugenics; the Kenya eugenic demographic fear was about African urbanisation and cultural change, not rapid population expansion. Finally, the concept of heredity was at the core of the Kenyan movement as an explanation of racial difference, but the Kenyan eugenicists were less interested in debates on the mechanisms of heredity than in examining biological evidence of innate racial difference. This reflected the fact that the most influential Kenyan eugenicists were medical doctors; the European population of the colony did not include geneticists or experts on evolution with the knowledge or professional interest necessary to contribute seriously to such questions. But, as we shall see in the following chapters, a lack of expert knowledge did not prevent Kenya's eugenicists from forming opinions or expressing their views about the African brain.

Notes

1 H. L. Gordon, 'Amentia in the East African', *Eugenics Review*, 25/4 (1934), pp. 223–35, 235.

2 D. Kevles, *In the Name of Eugenics: Genetics and the Uses of Human Heredity* (Cambridge, 1995), p. 59.

3 Ibid., p. 60.

4 See D. MacKenzie, 'Eugenics in Britain', *Social Studies of Science*, 6 (1976), pp. 499–532; 'Karl Pearson and the Professional Middle Class', *Annals of Science*, 36 (1979), pp. 125–43; and 'Sociobiologies in Competition: The Biometrician–Mendelian Debate' in C. Webster (ed.), *Biology, Medicine and Society, 1840–1940* (Cambridge, 1981), pp. 243–88. See also L. J. Ray, 'Eugenics, Mental Deficiency and Fabian Socialism between the Wars', *Oxford Review of Education*, 9/3 (1983), pp. 213–22. On links between Fabianism and Imperialism see E. Hobsbawm, 'The Fabians Reconsidered' in his *Labouring Men: Studies in the History of Labour* (London, 1968), pp. 241–71.

5 M. Freeden, 'Eugenics and Progressives: A Study in Ideological Affinity', *Historical Journal*, 22/3 (1979), pp. 645–71. See also D. B. Paul, 'Eugenics and the Left', *Journal of the History of Ideas*, 45/4 (1984), pp. 567–90. The debate between Jones and Freeden on this subject was continued in G. Jones, 'Eugenics and Social Policy between the Wars', *Historical Journal*, 25/3 (1982), pp. 717–28, and M. Freeden, 'Eugenics and Ideology', *Historical Journal*, 26/4 (1983), pp. 959–62. A useful, if slightly out-of-date, overview of the literature on eugenics in Britain is L. Farrall's 'The History of Eugenics: a Bibliographical Review', *Annals of Science*, 36 (1979), pp. 111–23, and more recently,

Dikötter's 'Race Culture: Recent Perspectives on the History of Eugenics', *American Historical Review*, 103/2 (1998), pp. 467–78.

6 Jones, 'Eugenics and Social Policy'.

7 See G. R. Searle, 'Eugenics and Class' in Webster (ed.), *Biology, Medicine and Society*.

8 P. Mazumdar, *Eugenics, Human Genetics and Human Failings: The Eugenics Society, Its Sources and Its critics in Britain* (London, 1992), p. 1.

9 Ibid., p. 2.

10 G. R. Searle, 'Eugenics: The Early Years' in R. Peel (ed.), *Essays in the History of Eugenics* (London, 1998), pp. 20–35.

11 Royal Commission on the Care and Control of the Feeble-Minded, *Minutes of Evidence and Reports*, Cmd 4215–4221 (London, 1908).

12 Kevles, *In the Name of Eugenics*, p. 99.

13 Mazumdar's *Eugenics, Human Genetics and Human Failings* provides a useful description of the aims and methods of the Eugenics Society.

14 Kevles, *In the Name of Eugenics*, p. 170.

15 R. A. Soloway, 'From Mainline to Reform Eugenics – Leonard Darwin and C. P. Blacker' in Peel (ed.), *Essays in the History of Eugenics*, pp. 52–80, 54.

16 Ibid., p. 53.

17 Ibid., p. 55.

18 Ibid., p. 63.

19 G. R. Searle, 'Eugenics and Politics in Britain in the 1930s', *Annals of Science*, 36 (1979), pp. 159–69, 159.

20 Ibid., p. 160.

21 Ibid., p. 159.

22 E. J. Lidbetter, 'The Social Problem Group as Illustrated by a Series of East London Pedigrees', *Eugenics Review*, 24/1 (1932), p. 9.

23 Mazumdar, *Eugenics, Human Genetics and Human Failings*, p. 139.

24 Kevles, *In the Name of Eugenics*, p. 176.

25 Mazumdar, *Eugenics, Human Genetics and Human Failings*, p. 211.

26 Ibid., p. 196.

27 Ibid., pp. 196–230.

28 *Report of the Mental Deficiency Committee being a Joint Committee of the Board of Education and the Board of Control (Wood Report)* (London, 1929).

29 'Report of Mental Deficiency Committee', *BMJ* (1932), Supplement, 25 June, pp. 563–77.

30 Board of Control, Committee of Sterilisation, *Report (Brock Report)*, Cmd 4485 (London, 1934).

31 L. S. Penrose, *A Clinical and Genetic Study of 1280 Cases of Mental Defect (Colchester Survey)*, No. 229 in Special Report series (London, 1938).

32 Kevles, *In the Name of Eugenics*, pp. 150–1.

33 L. S Penrose, *Mental Defect* (London, 1933).

34 L. S. Penrose, 'The Complex Determinants of Amentia', *Eugenics Review*, 26/2 (1934), pp. 121–6.

35 Mazumdar, *Eugenics, Human Genetics and Human Failings*, p. 226.

36 Penrose, *Colchester Survey*.

37 Kevles, *In the Name of Eugenics*, p. 159.

38 See D. Pick, *Faces of Degeneration: A European Disorder, c. 1848–c.1918* (Cambridge, 1989) and R. A. Soloway, *Demography and Degeneration: Eugenics and the Declining Birthrate in Twentieth-Century Britain* (Chapel Hill, 1990).

39 For example, see F. H. Hankins, 'Civilization and Fertility: Has the Reproductive Power of Western Peoples Declined?', *Eugenics Review*, 23/2 (1931), pp. 145–50, and F .C. S. Schiller, 'Eugenics versus Civilization', *Eugenics Review*, 13/2 (1921), pp. 381–93.

40 For example, see K. B. Aikman, 'Race Mixture', *Eugenics Review*, 25/3 (1933), pp. 161–6.

41 M. Banton, *Racial Theories* (Cambridge, 1987), p. 77.

42 N. Stepan, *The Idea of Race in Science: Great Britain 1800–1960* (London, 1982), p. 111. See also E. Barkan, *The Retreat of Scientific Racism: Changing Concepts of Race*

in Britain and the United States between the World Wars (Cambridge, 1992), p. 2; Banton, *Racial Theories*; J. Halliday, *Darwinism, Biology and Race* (Working Paper no. 49, University of Warwick, 1990); and K. Malik, *The Meaning of Race: Race, History and Culture in Western Society* (Basingstoke, 1996).

43 P. Rich, *Race and Empire in British Politics* (Cambridge, 1990), p. 206.

44 J. M. MacKenzie, *Propaganda and Empire: The Manipulation of British Public Opinion, 1880–1960* (Manchester, 1986), p. 7.

45 Ibid., p. 11.

46 J. M. MacKenzie (ed.), *Imperialism and Popular Culture* (Manchester, 1986).

47 R. Semmel, *Imperialism and Social Reform: English Social-Imperial Thought, 1895–1914* (London, 1960).

48 Aikman, 'Race Mixture', p. 166.

49 S. Dubow, 'Race, Civilisation and Culture: The Elaboration of Segregationist Discourse in the Inter-War Years' in S. Marks and S. Trapido (eds), *The Politics of Race, Class and Nationalism in Twentieth Century South Africa* (London, 1987), pp. 71–94, 74.

50 K. Pearson, *Social Problems: Their Treatment, Past, Present and Future* (London, 1912), p. 7.

51 K. Pearson, *The Function of Science in the Modern State* (Cambridge, 1919), p. 5.

52 Semmel, *Imperialism and Social Reform*, p. 28

53 L. Farrall, *The Origins and Growth of the English Eugenics Movement, 1865–1925* (New York, 1985), pp. 306–7.

54 W. C. D. and C. D. Whetham, 'The Influence of Race on History' in *Problems in Eugenics – Papers Communicated to the First International Eugenics Congress* (London, 1912), p. 243

55 C. W. Saleeby, *Parenthood and Race Culture* (London, 1909), p. 266. Saleeby's italics.

56 'Galton's View on Race', Appendix 1 in C. P. Blacker, *Eugenics: Galton and After* (London, 1952), pp. 323–8.

57 Barkan, *The Retreat of Scientific Racism*, p. 185.

58 Cited by Barkan, *The Retreat of Scientific Racism*, p. 178. Quotation from J. Huxley, 'America Revisited. III. The Negro Problem' in the *Spectator*, 29 November 1924.

59 J. Huxley, *Africa View* (London, 1931), p. 388.

60 Barkan, *The Retreat of Scientific Racism*, pp. 241–2.

61 J. Huxley and A. C. Haddon, *We Europeans: A Survey of 'Racial Problems'* (London, 1935).

62 On the role of race in anthropology, and Elliot Smith and Arthur Keith, see Barkan, *The Retreat of Scientific Racism*, pp. 15–65, and Stepan, *The Idea of Race in Science*, pp. 83–110.

63 Barkan, *The Retreat of Scientific Racism*, pp. 46, 65.

64 Ibid., p. 65.

65 Gordon to Blacker, 28 July 1938, CMAC, SA/EUG/C.130.

66 An important comparative study of eugenics internationally can be found in M. B. Adams (ed.), *The Wellborn Science: Eugenics in Germany, France, Brazil and Russia* (Oxford, 1990). See also A. Drouard, 'Eugenics in France and Scandinavia: Two Case Studies' in Peel, *Essays in the History of Eugenics*, and N. Stepan, *'The Hour of Eugenics': Race Gender and Nation in Latin America* (Ithaca, 1991).

67 L. R. Graham, 'Science and Values: The Eugenics Movement in Germany and Russia in the 1920s', *American Historical Review*, 83/5 (1978), pp. 1135–64.

68 N. Stepan, 'Eugenics in Brazil' in Adams (ed.), *The Wellborn Science*, pp. 110–52.

69 M. B. Adams, 'Towards a Comparative History of Eugenics' in Adams (ed.), *The Wellborn Science*, pp. 217–31.

70 L. Darwin, 'Eugenics and Imperial Development', *Eugenics Review*, 11/3 (1919), pp. 124–35.

71 Soloway, *Demography and Degeneration*, p. 60.

72 C. S. Stock, 'The Sex Ratio and Emigration', *Eugenics Review*, 10/3 (1918), pp. 163–6.

73 H. Lundborg, 'The Danger of Degeneracy', *Eugenics Review*, 13/4 (1921), pp. 531–9, 532.

74 Stepan, *The Idea of Race in Science*, p. 112.
75 S. Fazan, *An Examination of the Rate of Population Increase of the Kikuyu Tribe* (1932), KNA, VQ/1/20.
76 Kenya Colony and Protectorate Census, 1948, KNA, CS/1/1/3.
77 M. G. Redley, 'The Politics of a Predicament: The White Community in Kenya 1918–1932' (PhD thesis, Cambridge, 1976), p. 3.
78 Kenya Colony and Protectorate Census, 1948, KNA, CS/1/1/3.
79 D. M. Anderson, 'Depression, Dustbowl, Demography, and Drought: The Colonial State and Soil Conservation in East Africa during the 1930s', *African Affairs*, 83/332 (1984), pp. 321–43, 328.
80 H. L. Gordon, *The Population Problem in a Crown Colony*, 24 May 1945, Public Records Office, London (NA), CO 533/537/13.
81 A. Burton, 'Wahuni (The Undesirables): African Urbanisation, Crime and Colonial Order in Dar es Salaam, 1919–1961' (PhD thesis, London, 2000), p. 27. A valuable comparative discussion on urbanisation and the urban poor can be found in Chapter 1 of Burton's thesis, entitled 'Urbanisation in Comparative Perspective', pp. 24–35.
82 S. Dubow, *Scientific Racism in Modern South Africa* (Cambridge, 1995), p. 284.
83 Redley, 'The Politics of a Predicament', p. 12.
84 Ibid.
85 D. Kennedy, *Islands of White: Settler Society and Culture in Kenya and Southern Rhodesia, 1890–1939* (Durham, 1987), p. 58.
86 Ibid., p. 5.
87 Ibid.
88 P. Mosley, *The Settler Economies: Studies in the Economic and Social History of Kenya and Southern Rhodesia, 1900–1963* (Cambridge, 1983). Focusing just on Southern Rhodesia, G. Arrghi's 'The Political Economy of Rhodesia' in G. Arrighi and J. S. Saul (eds), *Essays on the Political Economy of Africa* (New York, 1973), pp. 336–77, discussed the differing class interests of Europeans, which were at times lessened by the formation of a coalition against international capitalism.
89 Mosley, *The Settler Economies*, p. 235.
90 Redley, 'The Politics of a Predicament', p. 14. Another useful source on settler politics is G. Bennett's 'Settlers and Politics in Kenya up to 1945' in V. Harlow and E. M. Chilver (eds), *Oxford History of East Africa* (Oxford, 1965).
91 Kennedy, *Islands of White*, p. 190.
92 C. J. D. Duder and C. P. Youé, 'Paice's Place: Race and Politics in Nanyuki District, Kenya, in the 1930s', *African Affairs*, 93 (1994), pp. 253–78.
93 Kennedy, *Islands of White*, p. 184.
94 Ibid., p. 278. Duder and Youé's perspective can also be contrasted with E. Huxley's *White Man's Country: Lord Delamere and the Making of Kenya*, Volumes I and II (London, 1935).
95 Dubow, *Scientific Racism in Modern South Africa*, p. 171.
96 Ibid.
97 Ibid., p. 170.
98 Ibid., and S. M. Klausen, *Race, Maternity, and the Politics of Birth Control in South Africa, 1910 –39* (Basingstoke, 2004), p. 43.
99 This detail on the eugenics movement in South Africa comes from Klausen's invaluable work in the area: S. Klausen, 'The Race Welfare Society: Eugenics and Birth Control in Johannesburg, 1930–40' in S. Dubow, *Science and Society in Modern South Africa* (Manchester, 2000), Klausen, *Race, Maternity, and the Politics of Birth Control*, and '"For the Sake of the Race": Eugenic Discourses of Feeblemindedness and Motherhood in the South African Medical Record, 1903–1926', *Journal of Southern African Studies*, 23/1 (1997), pp. 27–50.
100 Klausen, 'The Race Welfare Society', p. 165.
101 M. Vaughan, *Curing their Ills – Colonial Power and African Illness* (Stanford, 1991), p. 109.
102 M. Hailey, *An African Survey* (London, 1938).

103 V. Berry (ed.), *The Culwick Papers, 1934–1944: Population, Food and Health in Colonial Tanganyika* (London, 1994) and V. Berry and C. Petty (eds), *The Nyasaland Survey Paper, 1938–1943: Agriculture, Food and Health* (London, 1992).
104 Berry and Petty (eds), *The Nyasaland Survey Paper*, p. 2.
105 P. Hetherington, *British Paternalism and Africa, 1920–1940* (London, 1978).
106 Editorial, *EAS*, 15 July 1933, p. 42.

CHAPTER 3

Kenyan medical discourse and eugenics

In January 1931, Dr H. L. Gordon, President of the Kenya branch of the British Medical Association, made a speech at the organisation's Annual Dinner which was a powerful plea for the use of eugenics in colonial development policy. He argued that the promotion of education and physical health in Africa were potentially irresponsible objectives if undertaken without due regard for immutable African capabilities, or lack of them:

> Native backwardness could never be made to disappear under the mere trappings of civilization. No prevention of disease, bonification, education or religion could enable them to gather grapes off thorns or figs off thistles.[1]

Gordon went on to argue that:

> The practical corollary of those indisputable axioms is this, for both science and politics: As humane custodians of our precious civilization we are alive to the ghastly dangers of admitting immigrants merely on certificates written by their dear friends or prejudicial enemies. Determined as we are to help all worthy immigrants with our knowledge and power we are equally determined first to examine them and their natural passports for ourselves. The motto is 'Safety First'.[2]

In essence, Gordon was arguing that without a proper understanding of eugenics (and the eugenic limitations of Africans) the pursuit of African development was a foolhardy project. In the 1930s eugenics became a major preoccupation for the medical profession in Kenya and the Kenyan eugenicist doctors made their agenda central to debates about African welfare and development and related medico-legal questions.

In this chapter the theories of the predominant eugenicists will be analysed. Eugenics in Kenya grew out of the theories disseminated from Britain; the application of current ideas about the transmission of innate characteristics, in particular intelligence, shaped a new and extreme eugenic interpretation of racial difference. The Kenyan eugenicists did

[39]

not, however, use the most obvious methods, such as pedigrees, statistics and intelligence testing, which were applied by British eugenicists when assessing the intelligence of large social groups. When examining race, an area in which British eugenics had not prescribed a methodology, the Kenyan doctors most radically made histological counts of brain cells and physical measurements of brain capacity. This led to the adoption of a particularly pathologising theory about biological inferiority in the East African brain.

Kenyan eugenics largely originated from within the medical profession, which explains the eugenicists' choice of anatomical methods. It has been suggested that the eugenics movement in Britain was most actively supported by scientists who were working in relatively young areas, such as statistics and genetics, and were struggling to obtain for their subject credibility and respect from a potentially sceptical scientific or medical establishment.[3] This was not the situation in Kenya: the eugenic case was widely accepted within the local medical establishment. The cohesive role of the medical profession in Kenyan eugenics was partly a result of its small size and lack of specialists: there were fifty Medical Officers in Kenya in 1939 according to the Medical Service list.[4] The other important factor in this cohesion was the content of Kenyan eugenic racial thinking and its general confluence with a wider European consensus, in particular about race. The success of these ideas also needs to be placed in the context of the acceptance of eugenics generally by a large part of the medical administration in Kenya: for example, one doctor described the roles of a medical officer as being 'a clinician, a lecturer in eugenics, a Medical Officer of Health, an educationalist and an amateur agriculturalist'.[5]

Eugenic research on the African frontier: Dr Gordon and Dr Vint

This section will deal with the theories and researches of Dr H. L. Gordon and Dr F. W. Vint that became connected with eugenics, in particular their contributions to the subject of race and intelligence. The work of these two doctors, a psychiatrist and pathologist, was central to the articulation of the Kenyan racial theories. Gordon and Vint worked together in developing their ideas and dovetailed their research, but their roles in the formation of the Kenyan eugenics movement were quite different. Gordon was a propagandist, keen to educate Kenyan settlers in eugenics and establish a general eugenic programme of 'scientific colonization'.[6]

Vint was more reticent about making social and political extrapolations, maintaining a position of scientific detachment. Vint's role, although less vocal, was vital to the significance that became attached to the Kenyan research. He acquired a reputation as a serious scientist of excellent standards, influenced only by facts based on careful investigation and unswayed by political debate. The profound flaws and biased motivation behind Vint's work that strike a present-day observer should not obscure the fact that in his day, Vint was considered a talented researcher and an important asset to the advancement of science in Kenya, as was shown by protests made in 1936 when it looked as if he might be moved to Mauritius.[7] The potency of Vint's reputation as a serious scientist did much to win over support for Kenyan eugenics both in the colony and in Britain.

Dr Gordon, and to a lesser extent Dr Vint, have been discussed by historians before; the extremity and controversy of their theories on race and intelligence have earned their research particular notoriety. The infamy of Gordon and Vint has tended to result in their being considered outside their medical and settler contexts, which were in fact not only complicit, but also enthusiastic and highly supportive of their work. Essential to this context is the role of eugenic thought and scientific racism in Kenyan discourses on human biology and its social implications. McCulloch has discussed Gordon in some detail in the book *Colonial Psychiatry and 'the African Mind'*, placing him in the tradition of colonial psychiatry in Africa, largely because of the parallels that can be made between Gordon and the notorious post-war Kenyan psychiatrist, James Carothers.[8] One of the results of this approach has been to alienate Gordon from the professional context within which he was working and developing his theories.

The problem with focusing on a discipline, like psychiatry, when examining colonial medical history lies in the fact that there was much less specialisation and hence there were fewer boundaries between specialists because of the small number of doctors and scientists in a territory. The majority of doctors, both private practitioners and medical officers, worked in general practice and hence the discourse that developed within Kenya was shaped by miscellaneous interests and interdisciplinary connections. McCulloch has asserted that Gordon and Vint were distinctive in their use of physiology to explain racial differences.[9] However, to grasp the full influence of eugenic thought or scientific racism, it is necessary to examine a wider medical discourse that did not exclusively include psychiatrists; after all Vint was a pathologist. The very fact that Gordon and Vint's researches became so connected indicates that situating such research in one discipline is misleading. Similarly, in Dubow's study of scientific racism in South Africa, he has found a strong body of eugenic thought not just within the medical profession, but also among academics like Harold Fantham, Professor of Zoology and Comparative Anatomy at the University of Witswatersrand, and the criminologists at Pretoria, Willemse and Cronjé.[10]

Gordon's research was certainly not aimed at an audience of psychiatrists, but a wider constituency of scientific experts of varying disciplines who had an interest in race. The work of Dubow and Klausen and this study of eugenics in Kenya indicate that eugenic thought did not follow neat disciplinary boundaries, but spread between like-minded individuals. The history of an idea is complicated by the indirect processes of diffusion; this is perhaps even more true in the case of colonial territories where there were fewer institutional and academic confines. By reading Dr Gordon's articles, speeches and letters more extensively, it has become clear that he was well-supported in scientific and medical circles in Kenya, that the basis of his thinking lay less in psychiatry than in eugenics and biology, and that his intellectual partnership with Vint did more to shape his work on race than psychological theories. By 1934 Gordon looked back at his attempts at analysing backwardness through the discipline of psychology as fruitless, as conveyed in his description of his early approach in Kenya around 1926: 'I was speculating about the causes of backwardness and making futile excursions into native psychology.'[11] Gordon's ideas about race and intelligence were part of a more general eugenic programme for Kenyan development, including policies on Europeans: controlling immigration according to eugenic standards and preventing mental deficiency among Europeans through sterilisation and positive eugenics. It was, however, Gordon's stand on the issue of race and intelligence that gained him the greatest publicity, partly through his determined campaigning, as well as the appeal that scientific racism had through its complicity with settler racial hostilities.

Henry Laing Gordon (normally known as Harry) was born in Cumbria and received his medical training at Edinburgh University; he failed his finals at his first attempt and won a Gold Medal at the second. Owing to ill-health, Gordon was advised to live in a warm climate. He moved to Florence, where he met his second wife, with whom he eloped. Gordon travelled extensively, spending time in France, South Africa, Canada and most of World War One in Portugal. During this period Gordon tried his hand at different occupations such as apple farming and writing plays. After the war Gordon returned to London, where he worked as Physician to the London Neurological Clinic, Mental Specialist to the Ministry of Pensions and Assistant Physician with charge of out-patients at the West End Hospital for Nervous Diseases.[12] In London Gordon visited the British Empire Exhibition of 1924, which inspired him to move to Kenya.[13]

Gordon arrived in Kenya in 1925 and acquired a medical farm thanks to the scheme whereby private doctors were given cheap farms in order to get them into the country, providing greater access to medical facilities for settlers. Gordon was given a farm at Koru. In 1927 the Koru Farmers and Planters Association complained to the Provincial Commissioner about his medical practice.[14] One of the complaints was that he refused to supply himself with a car, and so often could not make calls on patients. A neighbour even offered to lend Gordon a driver and car, but this offer was declined. What is more, Gordon seems to have generally been rather reluctant to make visits; when prevailed upon, he would often refuse to attend a patient unless he was fetched by car, rather than ride, or he would refuse to visit because of his own indisposition. The other major complaint against Gordon was his medical practice for Africans. It was protested that when European farmers requested Gordon to treat a sick farm worker, Gordon would charge excessive fees where it was customary to charge lower fees for medical services, mainly because farmers would tend not to obtain treatment for an African worker unless it was cheap. Gordon seems to have wanted to avoid treating Africans; when a European farmer complained about the excessive charge Gordon made for treating one of his workers, Gordon agreed that it was expensive and said he did not expect the farmer to come back with another African patient.[15] Gordon found lime on his farm, which he acquired a licence to mine; he sold the farm to a lime-burning company and moved to the Nairobi area, leaving behind disgruntled local settlers who had lost their doctor and the medical farm that could have been used to attract another.

In Nairobi Gordon ran a private practice specialising in mental health.[16] In 1929 he was employed as a private consultant for European cases at Mathari Mental Hospital, the only mental health facility in the colony at the time.[17] In 1931 Gordon's role expanded to that of Visiting Physician, responsible for the psychiatric treatment of all races at Mathari.[18] Gordon remained at Mathari until 1937; as will be discussed in Chapter 6, he lost his position there because of his obstructive and tendentious line on mental health reform in Kenya. Although Gordon was a qualified medical doctor, he had no training in psychiatry or psychology. His role as Kenya's expert on psychology and his position at Mathari were founded on the fact that there was nobody else in the colony who knew more about these areas. This lack of qualifications did not prevent Gordon from achieving a high status as an expert on matters relating to the brain and psychology in Kenya; he was an active and well thought-of figure within the medical community. Flood (Assistant Secretary of State for the Colonies) described Gordon as 'an alienist [psychiatrist] of high repute and is Honorary Consultant to the Governor for Mental Diseases'.[19]

Gordon's standing within the medical profession was demonstrated when he became President of the Kenya branch of the BMA in 1931,[20] shortly after publishing an article entitled 'A Note on the Diagnosis of Amentia in Africans'. In this article he argued that there was a high level of inherited, innate mental deficiency, or amentia, in Kenya, which caused inferior intelligence in the 'East African native' and which a programme of eugenics could solve.[21] Gordon kept up a regular flow of articles in the *East African Medical Journal* (*EAMJ*): sixteen articles between 1928 and 1945, as well as fairly regular contributions in the form of correspondence.[22] Some of Gordon's writing, as well as his public lectures, were written from a eugenic point of view on issues such as the hereditary qualities of European settlers and sexuality, but the most crucial subject for him was the innate mental capacity of the 'East African native'. This led to a campaign that involved leading members of the Kenyan medical profession and of the British Eugenics Society for substantial investment from the British government in research into the intelligence of the East African. The aim was to establish Nairobi as a major research centre for analysing the causes of racial backwardness. Resistance in the metropole eventually meant that this goal was never realised, although Gordon continued publishing articles in the *EAMJ* on eugenics.[23] He died in 1947.[24]

From a eugenic perspective, research into African mental capacity held an allure based on the notion that Africa had been unaffected by civilisation. This led to the idea that the hereditary qualities of African people were of universal scientific interest, that Africa as an untouched continent was the ultimate control experiment when examining human behaviour. In his first letter contacting the British Eugenics Society, Gordon used this argument to emphasise the value of his research:

> You will appreciate the point that I am working on entirely virgin soil … I am convinced that this question of amentia among natives is likely to take an important place ultimately in relation to many African problems, and that because the soil is virgin conclusions may be reached of value to the science of Genetics generally.[25]

Gordon believed that funds should urgently be invested in examining the mental capacity of the East African before 'civilization queers the pitch'.[26] Gordon's conclusion was that a eugenic study of the inherited mental abilities of the race was not only of unique value to the advancement of science, but also that it would be of crucial importance in the management of empire. He believed that with the help of eugenics and an understanding of the histology of the brain of East Africans, correct policy could be empirically determined:

Cerebral deficiency affecting a population under British tutelage and not restricted to the types known in Europe as amentia – such deficiency has medical, legal, educational, and social aspects of Imperial importance.[27]

Henry Gordon first appears in the public discourse in 1926, when the *East African Standard* reported a lecture he had given on psychology and the African.[28] This early example of Gordon's thinking is revealing:

> Dr Gordon said that he would begin boldly and say that in spite of opinion to the contrary he had no hesitation in declaring that science was satisfied that the species, Kenya man, belonged to the genus, man, and had a mind![29]

This early article is unusual in that it does not mention eugenics: its language and approach were taken from psychology rather than biology. Instead of the terms 'nature' and 'nurture', commonly used in eugenic writing and used by Gordon himself later, he here used the more psychoanalytic terms 'Inborn Mind' and 'Acquired Mind'. The lecture is concerned with the same issue that would dominate his later work: differences between the African and European brain, but the emphasis was less on intelligence than on instinct, revealing a different intellectual frame of reference. Gordon described his abandonment of a psychoanalytic approach for a biological and eugenic one to Blacker of the British Eugenics Society:

> Until I came to Kenya six years ago I was content to treat Mind without reference to Brain. The whole gamut of Freud, Jung, McDougall etc had been mine, in study and practice. I fell back on brain – back to Shaw Bolton, Mott, Sherrington and so on; with reservations, I hugged Berry's 'Brain and Mind'; talked to the Government Statistician about Galton and Pearson ...[30]

Whereas it took several years for Gordon to elaborate eugenic thought for the purposes of racial study, he was obviously familiar with British eugenics, which he initially applied to the settler community. He wrote several articles on syphilis in the *EAMJ*, which reveal his interest in the social aspect of biological conditions, for example, 'The Social Aspect of the Curious History of Syphilis'.[31] In this he wrote with admiration of a colleague's belief that 'if we labour only for the individual medicine is little better than a trade; if we labour also for the race it is an inspiration'.[32]

Gordon argued in the article 'Mental Instability among Europeans in Kenya' that the colonial environment required a higher eugenic standard than the metropole and that there was a lack of rigorous testing for intelligence and sanity in the acceptance of new settlers into Kenya. He argued that Kenya was in danger of becoming a dumping-ground for the mentally unstable in Britain, particularly from among the upper classes:

> Our race does not get rid of the threat of deterioration by transferring some of the threat to where the Empire requires only the best. The following brief notes may emphasize that while Kenya takes great care to see that immigrants have the right kind of pockets she takes none to see that they have the right kind of minds.[33]

Gordon concluded by emphasising that mental instability of all degrees was present in Kenya's settler society, and that its source was from outside the colony itself, in the Europeans who were sent to Kenya to improve or conceal their condition. Although Gordon thought that some of these 'mental unstables' could be cured or improved sufficiently to become useful members of the colony, it was necessary to stop the influx of such dysgenic individuals.

A dramatic and urgent exposition of eugenics and its importance in the colonial environment was made in the pamphlet 'Eugenics and the Truth about Ourselves in Kenya'.[34] Based on a popular public lecture Gordon had given in Nairobi in March 1933,[35] it was printed by the *East African Standard* to assist the formation of the Race Improvement Society.[36] The pamphlet defined and explained the basic tenets of eugenics, emphasising the importance of heredity and stating that 'Eugenics aims to raise the average ability of the race at birth, by greater attention to nature'. The first way in which this was to be achieved was by getting 'rid of the breeding of BAD HUMAN STOCK', the second was through the encouragement of 'the best of our race' to have more children. Gordon warned that currently 'Our middle classes, the very backbone of our race are now producing only 40 per cent. of the numbers required to replace them … *If this is allowed to go on we cannot avoid the sinking to a level beyond rescue.'*[37] Gordon went on to apply mainline British eugenics to the racial circumstances of Kenya. The African population replaced the British, largely urban lower working classes as the main preoccupation about genetically inferior stock threatening to swamp the country:

> We are developing a *poor white group*, a *submerged Asiatic group,* and a *huge African group* of alarming potentialities. The reason is most evident in the case of the native.

> TRUSTEESHIP
> is being interpreted as *nurture* only. We are trying to create a new civilisation by repeating the old problem of neglecting nature.[38]

Gordon urged that in the face of these dangers, it was essential that only fertile, high-quality British stock be admitted to the colony, 'not retired breeders from Anglo-India or elsewhere'.[39]

The first piece of research that Gordon undertook on African intelligence was his study of mental deficiency at Kabete Reformatory.[40] In 1930 he produced a 'Report of a Survey of the Inmates of Kabete Reformatory for the Purpose of Detecting the Presence of Amentia (Mental Deficiency)'.[41] Gordon's work at the Reformatory was significant partly because the results were produced at a time when a serious review of the Reformatory and the management of juvenile offenders in Kenya was beginning, which will be discussed in more detail in Chapter 6. In terms of the development of Gordon's ideas, his research at Kabete was important because although it was not published, it became well publicised among officials and gave Gordon the credibility of original research. John Gilks, the Director of Medical Services, sent copies of Gordon's report to the Colonial Secretary, the Chief Justice and the Committee of Visitors to the Reformatory, which included the District Commissioner, the Chief Native Commissioner and the Senior Resident Magistrate.[42]

Gordon's experiment at the Reformatory used a range of different methods to come to the extraordinary conclusion that of the 219 inmates he measured, not one attained the European standard of normal intelligence. Gordon claimed to have found that 45 were medium-grade aments (imbeciles), 144 were high-grade aments (idiots), 24 were borderline normal and 6 were low-grade normal.[43] Gordon created these two categories of normality because he claimed he could find no result that fitted in with European standards of normal intelligence. This led to the idea that the normative standard of African intelligence was different and that the level of this different standard had to be established to test African intelligence fairly, as to project European normality would be unrealistic. Gordon considered this survey at the Reformatory as evidence that innate amentia was widespread among East Africans and suggested that its presence was particularly worrying from the eugenic point of view, as it normally went undetected:

> the majority are to be regarded as carriers of neuropathic defect who, away from the competitive and economic forces of civilisation will be able to live and reproduce their like with the well-known vigour and success of the ament.[44]

The idea for this survey originated with Dr Kauntze, and was read and supported by Dr Gilks, Director of Medical and Sanitary Services in Kenya; both these doctors, as we shall see in more detail later, became important allies of Gordon in his campaign for eugenic research.

In the course of his inquiry, Gordon made over forty visits to the Reformatory and used a combination of measurements and tests:

Anthropological measurements, of physiological tests, of clinical examination more particularly of the nervous system and to detect abnormalities
attributable to defects of germ plasm (so-called stigmata), and of psychological or mental tests.[45]

Gordon borrowed this combination of methods from the English
anatomist and eugenicist, Professor R. J. A. Berry, who believed that
amentia was almost always caused by inadequate brain development
resulting from hereditary failings.[46] Berry, FRCS, FRSE, was the Director
of Medical Services at the Stoke Park Colony in Bristol. This was an institution established under the auspices of the 1913 Mental Deficiency Act;
it was designed to provide the accommodation in which to segregate the
mentally deficient under the Act.[47] By 1933 Stoke Park had about 1,700
patients, on whom Berry performed research to test his theories about
amentia. Berry was a traditional, 'mainline' eugenicist, whose work (as
seen in the previous chapter) was eventually undermined by the new
approach of Penrose's Colchester Survey. Berry was certain that disease,
accident or environment did not cause mental deficiency, but hereditary
flaws that could only be solved by preventive medicine in the form of
eugenics and embryology.[48]

Berry and Gordon were in contact with each other: Berry referred to
Gordon and Vint's researches in Kenya as an example of the application
of the methods that he had developed.[49] This shared methodology
included measurements of the cubic capacity of the brain, as well as more
general physical measurements. An abstract of a talk given by Gordon to
the Kenyan BMA was included in a book edited by Berry on mental deficiency. The talk was not on the subject of native backwardness, but on a
highly publicised trial in Kenya of a young British man called Ross who
murdered two women and was eventually hanged for it.[50] Gordon acted
as the psychological adviser for the defence and argued that the young
man was a high-grade hereditary mental defective and as such should not
be held responsible for his actions. Gordon went on to argue that the
blame for the crime rested with society as Ross should have been diagnosed and controlled, and ideally, in a society where eugenics was prac-
tised, should never have been born. Berry strongly endorsed Gordon's
interpretation of the crime and the way to tackle it.[51] The inclusion of
Gordon's interpretation of the Ross case in Berry's book is an example of
how eugenic thought in Kenya made its way back to the metropole and
how Gordon's applications of eugenics interested British eugenicists.

In Gordon's article 'A Note on the Diagnosis of Amentia (Mental
Deficiency) in Africans', further indication of his influences was given
through a series of quotations at the beginning of the piece. These include
Professor Shaw Bolton, who was Professor of Mental Disease at the

University of Leeds. Shaw Bolton wrote an influential book entitled *The Brain in Health and Disease*, in which he described amentia as a form of mental deficiency that was innate, arguing that it was 'in the widest sense *the mental condition of patients suffering from deficient neuronic development*'.[52] This definition of amentia was quoted by Gordon.[53] Tredgold's book, *Mental Deficiency (Amentia)*, the major textbook on the subject, was also cited by Gordon. It was first published in 1910 and remained widely read for a long time, with numerous editions.[54] Tredgold feared that degeneration was occurring in the British population, largely because of the dysgenic effects of town life; he approached mental deficiency from a eugenic perspective, believing that about 80 per cent of amentia was caused by hereditary factors. Gordon also included a quotation from Berry in which he warned of the hundreds of undiagnosed cases of amentia that were caused by a hereditary failure to produce sufficiently developed brain cells.[55]

The relationship between Gordon's work and that of figures such as Berry and Shaw Bolton is significant because it indicates that although Gordon's racial theories appear to be a form of outlandish and crankish pseudo-science, they were shaped by theories and methodologies being applied by experts in the metropole. The major difference between Gordon and Berry's research was that Gordon was working on a racial scale, whereas Berry was testing individuals diagnosed (albeit dubiously in the case of unmarried mothers, for example) as mentally deficient. Although eugenicists in Britain rarely showed particular interest in race, it is interesting to see how in Kenya, British eugenic thinking was transplanted and applied racially. Berry actually thought that certain physical measurements of intelligence, like brain weight and capacity, were more meaningful and accurate on a racial and class level than on an individual level.[56]

In 'A Note on the Diagnosis of Amentia in the East African' Gordon suggested that the failure to diagnose large numbers of high-grade aments in Kenya caused native backwardness:

> free breeding of high grade aments under primitive conditions may be a hidden eugenic factor in racial retardation in Africa, just as it may be (and is believed to be) a grave threat of racial degeneration in Europe. It is conceivable that absence of economic pressure and of industrial competition, coupled with advantages of tribal life in Reserves, leaves unrestricted the fecundity of high grade aments.[57]

High-grade amentia was the least severe form of amentia; it implied feeble-mindedness rather than idiocy or imbecility. A high-grade ament was considered more or less capable of self-support through work and the condition in adulthood was generally characterised by 'social' defects:

> The range of defects in this class is great, and the borderline between it and the normal may be difficult to distinguish. Cases occur which while the social defects are marked and permanent the intellectual abilities rise above the average normal and even into 'genius'. This, the most numerous class is the most disturbing and potentially dangerous to society; the least easy to detect and place under control, the most easy to conceal in the family.[58]

The diagnosis of mental defect through social judgments provided great scope for the diagnosis of Africans as high-grade aments who performed well in intelligence tests, but failed to meet Gordon's social criteria for normality. Cultural differences, linguistic misunderstandings and bias meant that Gordon was clearly not in an ideal position to assess the nuances of social normality in the black Kenyans he encountered.

The sheer volume of amentia that Gordon believed to exist in the African population changed the standards that could be expected, or as Gordon put it, it posed the question: 'When is a native an ament?'[59] Gordon went on to quote research he had conducted on 240 male Africans, aged from ten to nineteen. This research involved mental testing and physical measurements of 'mental equipment'.[60] Unfortunately Gordon did not give the results of this research or indicate where it was undertaken. Given the age of the subjects it is likely that the work was carried out at an institution such as a school or the Reformatory. The research is similar to that performed at Kabete Reformatory both in age range and in numbers, although Gordon makes no reference to his report on the Reformatory. Such vagueness about data is a recurring problem in Gordon's work.

The hypothesis put forward in 'A Note on the Diagnosis of Amentia in the East African' lays the basis of much of the Kenyan eugenic argument about the innate mental capacity of the native and the implications of this for social policy. The theory was that among 'raw natives' the cultural level was so low that high-grade aments could flourish. It was when the native was no longer in the reserve and introduced to 'civilisation', through urbanisation, European employment or education that the problems of amentia became clear. Thus, the problem of native amentia was clearly linked with that of development. The incidence of amentia was so high that it became the normative level within that population, with implications for policy regarding the entire African group; the numbers were too great to allow the application of traditional eugenic policies of segregation and sterilisation. The solution instead was to establish different standards for the entire African population that in effect meant segregation at a profound intellectual as well as practical level.

Gordon himself frequently reiterated his faith in scientific objectivity. This can be seen in his comments on the work of his colleague, Dr Vint:

Dr Vint has lifted us out of conflict and confusion into the serene and simple air of science. If we desire – and who does not? – that the native shall not be left misunderstood and helpless before inevitable new forces, then *wishful* thinking on all native questions must vanish from to-day before *scientific* thinking and mental hygiene must take its place as the premier public health problem in Kenya.[61]

Vint had qualified with a first from Belfast in 1924 and worked as a pathologist at the Royal Belfast Hospital and at Belfast University. He undertook post-graduate research at the London Hospital before moving to Kenya in the late 1920s.[62] Vint had a successful career in Kenya, frequently publishing in the *EAMJ*, being promoted to Senior Pathologist in the late 1930s and becoming President of the Kenya branch of the BMA in 1941.[63] Vint's professional respectability is indicated in his assumption of the role of medico-legal adviser to the government.[64]

Vint wrote two articles that were particularly pertinent to Gordon's hopes of scientific colonisation: 'A Preliminary Note on the Cell Content of the Prefrontal Cortex of the East African Native' and 'The Brain of the Kenya Native'.[65] It is interesting to observe how these key writings relate to the rest of Vint's research interests. Vint published the results of his work fairly consistently: thirteen of his articles were printed in the *EAMJ*, from his first in 1928 to 1945.[66] Of the articles based on his own research, the experiments usually involved post-mortems. The most interesting point about Vint's post-mortems is that they were all performed on Africans, usually adult males. He never appears to have been interested in performing post-mortems on Europeans for research purposes; European anatomy is usually mentioned in a comparative context, using other people's finding on Europeans as evidence. Vint published on many different subjects, including 'Gall Stones in a Native Child', 'Measurement of Red Blood Corpuscles' and 'Solar Rays. Fact or Fiction?'.[67] There was, however, a consistent pursuit throughout almost all of his work, which was to discover any racial differences in the physiology of Africans in comparison with Europeans. A typical example of this can be seen in his article, 'Measurement of Red Blood Corpuscles'. The objective of this experiment was to compare the size of the blood cells of Africans with those of Europeans. Vint concluded that the 'average size of the R.B.C of the natives of Kenya is .2 mu greater than that of the European R.B.C'.[68] Vint did not make any further racial inferences from such findings. Indeed, in many of his articles Vint suggested that the differences that he found should be attributed to environmental factors rather than any innate racial divergence. For example, in his article on the livers of East Africans he argued that European livers were usually smaller, and suggested that a high incidence of intestinal helminths (parasitic worms) in Africans might be the cause.[69] The underlying principle of Vint's research interests was that racial typologies were a matter of valid scientific interest and that answers to the problem of race could be found in human biology rather than a political or social interpretation.

Dr Gordon and Dr Vint were by no means unusual in their preoccupation with race; it was certainly a major interest of many of the doctors contributing to and, according to the correspondence, reading the *EAMJ*. The way in which this interest was expressed, however, was usually not explicit. As in much of Vint's work, it was indicated by the methodology and terms of reference. There was a tendency to differentiate racially when writing about physiology and also to segregate articles according to the race of the patients involved. Thus, many of the papers, although often written in apparently neutral terms, were concerned with illness in the African, or the European, rather than assuming a universal human biology. With titles like 'Rheumatic Fever in the Lumbwa Native' and 'The White Man in East Africa', of the forty-five articles in the subject index of the 1932–33 volume of the *EAMJ*, almost half are of this racially defined nature.[70] A quarter of the editorials of this volume of the *EAMJ* were concerned with aspects of racial differences in health from a biological as opposed to an environmental slant.[71]

The fact that practically all of Vint's work was concerned with race, and yet he avoided, except in his articles on the brain, the kind of dramatic generalisations that Dr Gordon was prone to, made him a more powerful ally in propagandising the supposed empirical reality of African mental inferiority. Thus, Vint's work was, for example, taken more seriously by the Colonial Office in London. As Flood said of Vint, 'He deals in observed facts'.[72] In fact, Vint was not disinterested when it came to the question of African intelligence. This can be seen in his correlation of brain size and weight with intelligence. The idea that such measurements could reliably reflect mental capacity and that a neat taxonomy of races could be made according to such a variable had been widely discredited. Indeed, such notions had been questioned as early as 1912 in Pearson's statistical journal, *Biometrika*, in an article written by Crewdson Benington and prepared for press by Pearson.[73] Yet Vint was unwilling to accept this:

> It is commonly believed that there is little or no relationship between the weight of the brain and 'intelligence' and in support of this is quoted the contrast between the large heavy brain of one class of mental deficient and the occasional small brain of a distinguished man. Physiologists and psychologists have tried and failed to establish a clear relationship between the size of the brain and 'intelligence,' but this failure does not preclude the possibility that pathology might succeed; in fact from the pathological standpoint weight of brain does appear to be related to 'intelligence'.[74]

Vint's ideas about brain weight were ventilated in his article 'A Preliminary Note on the Cell Content of the Prefrontal Cortex of the East African Native'. The argument that brain weight and intelligence were linked was a crux for Vint's theory that East Africans were less intelligent:

> Thus from the both the average weight of the African brain and measurements of its prefrontal cortex I have arrived, in this preliminary investigation, at the conclusion that the stage of mental development reached by the average native is that of the average European boy of between 7 and 8 years of age.[75]

For this research on the African brain there were two sections to Vint's argument: the first was concerning brain weight, the second was about brain structure, the measurement of its different parts and its cell content. Vint carried out post-mortems on adult male Africans in African hospitals in Nairobi. For the first part of the experiment Vint examined 351 brains, for the second part, only 35. He took no account of difference in tribe and does not seem to have made any selection on the basis of cause of death and did not have any information on the ante-mortem mental ability of the subjects.

According to Vint, the average weight of the African brain was 45 ounces, compared with just below 51 ounces for the European. Vint attributed this lesser weight to either a lack of nerve cells and the myelinisation of their axons or a lack of neuroglia tissue, or both. Inadequate numbers of nerve cells were thought to cause amentia, the function of myelin is to insulate axons, or nerve fibres – if there are fewer nerve cells there are fewer axons and consequently less myelin: 'Therefore there will not be the normal increase in the weight of the brain during post-natal development. This condition of affairs is found in amentia.'[76]

The question of brain weight was only part of Vint's argument. The other concerned his measurements of the prefrontal cortex. This part of the study involved measurements of the five layers of the prefrontal cortex of thirty-five adult Africans. Vint carried out these measurements at four different points of the brain and then took an average from these four points. Vint then compared his results with those of Von Economo and Shaw Bolton, who had both made similar experiments, although Bolton's results were used more seriously as a fair comparison because Economo's methods were unclear.[77] As with much of his work, Vint used the writings of Shaw Bolton on the brain as a basis for his methodology and his hypotheses. Given that Bolton's work was on Europeans, it was also used as a source for making racial comparisons. Taking all five layers together, the difference in average measurements between European and native was not very exciting to Vint. Shaw Bolton's total for Europeans was 1.89 mm and Vint's for Africans was 1.85 mm. To Vint the significance of these measurements lay in the discrepancy in the second layer, called the 'pyramidal cell layer':

> This is the last layer to be evolved and it is the first to undergo degeneration in mental disease and its instability is regarded as the basis of minor functional disorders. It is thinner in the lower animals and in the feeble-minded than in normal man. It presides over the highest functions as manifested by the capacity to learn from individual experience, and the power of learning.[78]

According to Table I in Vint's article, Shaw Bolton's measurement for the European second or pyramidal layer was 0.84 mm, while Vint's own measurement for the African second layer was 0.70 mm. In a minor inconsistency, however, Vint later gave Shaw Bolton's figure for the second layer as 0.83 mm.[79] From these figures, Vint argued that 'the pyramidal cell layer of the native brain has only attained to 84% of the development of the European'.[80]

Vint not only argued that the size of the second layer was different; he also emphasised his finding that the composition of the cells making up that layer was different. He found that in the African brain, there were a greater number of small undifferentiated cells and fewer large pyramidal cells than there were in the European brain. This had a bearing on arguments about intelligence because:

> The higher the mammal the more developed is this pyramidal cell layer. It is concerned with associational activities in that it associates and correlates the effects of previous stimuli, whence there is developed the powers of memory and deduction, in other words, mind.[81]

On the basis of comparisons with other animals, regardless of the fact that his sample consisted of only thirty-five brains and the well-established scientific evidence indicating that it was impossible both to categorise racially the brain and to correlate brain size with intelligence, Vint arrived at the radical conclusion that the average adult African attained the mental development of the seven- or eight-year-old European child. Vint tempered the extremity of his remarks by finishing off with the comment that it was impossible to say how much the immature nerve cells (the undifferentiated as opposed to the pyramidal cell in the prefrontal cortex) might develop under the influence of education and a change in environmental circumstances.[82]

Dr Gordon clearly had an important influence on Vint. At the beginning of his article on the prefrontal cortex, Vint stated: 'This investigation was undertaken at the suggestion of Dr H. L. Gordon.'[83] This paper was read before the Kenyan branch of the BMA, and the discussion that followed it, as well as the paper itself, was written up in the *EAMJ*. The discussion reveals the nature of the collaboration between Gordon and Vint. The importance of Vint's paper for Gordon is clear:

> If I have deprecated hasty criticism and conclusion I see no reason to moderate admiration for Vint's work. If I began as a feeble instructor to him I rose rapidly as a grateful pupil. In that capacity I take leave to congratulate him on having accomplished in 18 intensive months a certain end I have striven after without success for six years.[84]

Evidence of Vint's continuing commitment to Gordon's eugenic thinking can be seen in Vint's presence at many of Gordon's public lectures: for example Vint was in the Chair for Gordon's speech on 'Mental Degeneracy' in June 1939 at the Empire Theatre, Nairobi.[85]

Vint published his second work on the brain of the East African in the prestigious British *Journal of Anatomy* in 1934, in a paper entitled 'The Brain of the Kenya Native'.[86] This article gave the final results of his earlier research described in 'A Preliminary Note'. By this stage Vint had performed macroscopic and microscopic examinations on one hundred brains of Kenyan natives. The brains were taken from adult males obtained from the native hospitals of Nairobi. Cases from the Mental Hospital and prisons were rejected. Forty-eight of the subjects were Kikuyu, sixteen were Kavirondo, fifteen Jaluo, eleven Wakamba and ten were from other tribes.[87] In this paper Vint largely reiterated the findings given in his earlier paper: that the average weight of the Kenyan brains was less and the pyramidal cells of the supragranular cortex were smaller.

The article in the *Journal of Anatomy* was far more restrained than his earlier *EAMJ* article; intelligence, for example, was not directly mentioned at all. Vint still argued that the brain of the East African was smaller than that of the European, that both the cortex and the cells in the cortex were smaller, thus clearly making a comparison between the brain structures of Africans and Europeans, but he did not explicitly extrapolate inferior mental development from this.[88] A comparative approach to intelligence was clearly implicitly present, but it is interesting to note Vint's adoption of more temperate language when presenting his ideas to metropolitan scientists. The difference in approach indicates the difference between colony and metropole when it came to scientific attitudes towards race. The use of racial taxonomies in thinking about intelligence had not been entirely discredited within the British scientific community in the early 1930s: they were still considered a valid subject of scientific scrutiny, but the application of such typologies was much more inhibited, especially when it came to questions concerning such multifaceted areas as intelligence and mental development.

At the end of 1933 Gordon was given special leave by the Kenya medical department to allow him to go to London and test his ideas on the more specialised scientific opinion there. An indication of Gordon's standing is given by that fact that he was granted three months' paid leave to make this trip to London and his passage to Britain was also paid for by the Kenya government. Paterson (Director of Medical Services) argued that it was highly desirable that Gordon's research be communicated to a British audience, as the publicity would be valuable to Kenya and might lead to a serious research project.[89] Gordon was also to be paid expenses to enable him to visit the latest mental hospitals because in his capacity as Physician at Mathari he was advising the Kenya government on new lunacy legislation. During this visit to Britain Gordon presented a paper on the mental development of the African at a meeting of the Eugenics Society in London. He also read a paper on spirochaetal diseases before the Royal Society for Medicine and another paper on mental development before the Royal Anthropological Institute and gave an address to the African Circle at Chatham House.

The paper Gordon presented to the Eugenics Society was entitled 'Amentia in the East African' and was published in the *Eugenics Review* in 1934.[90] It reported on an inquiry Gordon had made in which he had taken the measurements of 3,444 'East African natives' recruited through the Native Labour Registration Office in Nairobi, by leave of the Chief Native Commissioner. The 3,444 subjects were all male and their ages ranged from nine years to adult; it was attempted to limit adults to those under the age of thirty as Gordon had a theory that the East African mind degenerated prematurely. Gordon and Vint expected senility in the East African at any time after the age of thirty-five, and attributed the lack of individuals over the age of sixty on the post-mortem table to this early brain decay, which was believed to set in after the age of eighteen.[91] The physical measurements that Gordon made were derived from Berry's method: cubic capacity of the brain estimated by head measurements, body weight, standing and sitting height, right and left grips, and vital capacity. The subjects were also given intelligence tests. Gordon's research was planned in conjunction with Vint's post-mortem research on brain structure.

Gordon reported that the average brain capacity of an adult male East African was 1,316 cubic centimetres, 165 cubic centimetres less than the European average. Gordon also compared the rate of growth of brain capacity of the East African and European:

> The European curve of 'actual' capacity begins, at the age of 10, almost where the native curve ends at the age of 20. Again, the steep ascent of the European curve from about the age of 14 makes a striking contrast to the

flat native curve. The curves show that between the ages of 10 and 20 the European average growth of brain is 17.7 cubic centimetres; the native yearly average is only 8.5 cubic centimetres.

Turning now to the curves of percentage volume increase we see that at the age of 10 the brain of the native is already 92.8 per cent. of its total volume.[92]

Gordon correlated this conclusion with the settler stereotype that the young 'native' boy before puberty tended to be:

a bright, malleable, nice little fellow, often – as one says – quick in the uptake; but that after puberty he almost invariably falls away from promise, and disappoints our hopes by lack of development in the very period where the European adolescent generally justifies hopes by rapid development.[93]

Another part of this research was a comparison of the brain capacity of different groups. These consisted of normal Europeans and four native groups: educated, mentally disordered, mentally deficient and criminal. Gordon predicted that the normal European would have the highest capacity, followed by the educated native, then the mentally disordered, the criminal and finally the aments. Gordon found his predictions were correct. The most significant result for Gordon was that even the educated native was found to have a lower brain capacity than the normal European. Gordon approached the question of environmental factors by arguing that the educated natives were of 'inherent cerebral superiority' to the other native groups, but still performed worse than the Europeans in the brain capacity measurement. The question of superiority within the African groups led Gordon to some conclusions about differences between tribes:

We see that the racial distribution of the aments corresponds to that of the 1,290 adults from whom the average native curve was obtained, i.e. the majority of them were Bantu; over 80 per cent. The distribution in the "educated" series is very different; 57 per cent. were from tribes commonly credited with inborn superiority to the Eastern Bantu and of these 48 per cent. were very superior Sudanese foreigners.[94]

Gordon also referred to Vint's work in this article, quoting Vint's conclusion that the total native inferiority in the quantity of cells in the prefrontal cortex was 14.8 per cent. Gordon concluded from this:

These, I think, are enough of Dr. Vint's new facts to make us feel that the deficiencies found in examination of the living are indeed associated with suggestive deficiency in the native cerebrum; that we are in fact confronted in the East African with a brain on a lower biological level.[95]

When Gordon presented this paper to the Eugenics Society he received a warm response.[96] This apparent acceptance of Gordon's methods by both his Kenyan colleagues and the British eugenicists is noteworthy given the fact that such crude correlations of head size and intelligence had been attacked by biologists and psychologists: the IQ test had become a more respected method of ascertaining intelligence.

Despite Gordon's emphasis on the objectivity of science in providing answers to the problem of racial backwardness, there are fundamental flaws in his research that call into question his scientific integrity. The results presented by Gordon to the British Eugenics Society were considered suspect by those involved in the African Research Survey in Oxford. Matheson, secretary of the Survey, wrote:

> It was of course obvious that his conclusions, even if substantiated, could be used by interested people to make all kinds of unwarrantable deductions, and I was rather worried about it. I had spoken to Julian Huxley, who was of course fully aware of the implications, and I have had a long talk with Mr Worthington who had already spent two hours with Dr. Gordon seeing all his graphs and other material; and while he thinks that he has got hold of some interesting material he is extremely suspicious of Dr Gordon's graphs; he says they are too perfect to convince anyone, and that no scientist ever gets results where everything appears to fit your theory so completely and nothing occurs to be placed on the other side.[97]

The doubt expressed by Worthington about Gordon's dubiously immaculate results brings us to a wider point about the thoroughness of Gordon's research. The research Gordon described in 'Amentia in the East African' tested an enormous subject group of 3,444 individuals and involved a large and time-consuming range of tests.[98] Gordon at that time had not received any outside funding for this labour-intensive research beyond a small grant of £50 awarded by the BMA.[99] Gordon was known to have little money himself to have paid for assistance in such a study,[100] which was partly why the Legislative Council gave him an honorarium of £100 on his retirement from Mathari.[101] At the same time that Gordon was conducting all this research, he was a visiting physician at Mathari Mental Hospital and running a private practice. Criticism of Gordon in a 'Report of the Board of Inquiry' into Mathari in 1932 casts further doubt over his reliability, in particular his organisational failings. There was also complaint about his lack of availability that echoed the earlier complaints made about Gordon's minimal work ethic when practising as a settler-doctor at Koru.[102] The overall picture of Gordon's circumstances, character and training suggest that he was not in a position to conduct reliably a research project of the scale he described. The research was planned in conjunction with Vint's post-mortem research on brain structure and the precision with which Gordon's results agreed with Vint's is a further cause for concern. It is impossible to substantiate the suggestion that Gordon was not entirely honest about his research. The raw data that he collected does not survive as he destroyed all his personal papers relating to the brain research before he died. Clearly Gordon did conduct research of some sort and was considered genuine by his contemporaries in Kenya, including high-ranking officials. It would appear, however, that with hindsight there is some cause to question the thoroughness of Gordon's methods and the reliability of his results, as those involved in the Oxford Research Survey did at the time.

Gordon's interpretation of the social applications of the theories on native backwardness became even more baldly asserted as his ideas reached a wider audience, as demonstrated in his article 'The Intentional Improvement of Backward Tribes' which he read before the KSSRI in July 1934. It was then read before the New Education Fellowship Conference in Johannesburg on 20 July, under the title 'Modern Knowledge and Education of Backward Tribes'.[103] By this stage, the campaign for research into African intelligence was well under way in Britain and Gordon had achieved the approval of some major British scientists and medics. Similarly, Dubow has described how Gordon's and Vint's work was widely reported and sympathetically received in South Africa.[104] Optimism and elation about the reception of this campaign are expressed in Gordon's South African talk:

> Most of you know of the proposal for a great scientific research into the causes of native backwardness and of the unanimous support it received in England last year from the most eminent in science. This year, again unanimously, it has gained scientific support all the way from Khartoum to Capetown. It has been my privilege to submit the proposal to Government in London and here and nothing has impressed me more than the sagacious caution of high officials towards a matter so new and so far-reaching and their repeated tributes to the impartiality of the scientific spirit ... Tonight I have come again into the whirligig of words on the subject to make a plain statement and I hope, to wipe away misunderstanding and prejudice from this momentous question while winning the co-operation of the few unbelievers left. I am fully convinced that honest understanding of the subject must give Kenya an unprecedented opportunity to cause the future to tell of her wisdom and show forth her praise; the opportunity to lead Africa into a new era of social justice and progress based for the first time in history on scientific knowledge.[105]

Throughout his writing, Gordon tempered some of the extremity of his theories by emphasising that scientific knowledge was at an early stage in the area of brain and racial backwardness and reiterating the need for large-scale scientific research to uncover the causes of African backwardness. He argued that while definite answers were still being awaited, the education and development of the African should be allowed to proceed slowly and with great caution. Indeed, he argued that it was in the best interests of the African to establish suitable levels of development and not to apply European levels, which might be inappropriate to African capacity:

> To what extent is it biologically and socially safe to urge upon one race the behaviour and institutions of another race if the two races differ biologically and socially? More specifically – are the ideals of life, liberty, and the pursuit of happiness, developed and cherished by an advanced people through 'the long drip of human tears' remembered, – are these necessarily the ideals for a backward people old also but in tragedy forgotten?
>
> Next, to what extent may the ideal of trusteeship safely regard the East African as a 'tabula rasa' and give us leave to scribble on it as we choose in many inco-ordinate ways?[106]

Despite such caveats, Gordon described some of the disastrous results of not taking into account the different standards of innate African mental capacity.

One consequence was elaborated in Gordon's paper read before the African Circle during his trip to London at the end of 1933. The paper was published in the *Journal of the African Society* in July 1934;[107] the journal (renamed *African Affairs* in 1944) which at that time served as a 'voice of the colonial system from which so many of its early contributors were drawn'.[108] Gordon's argument in this article concerned the condition dementia praecox (the old term for the symptoms now associated with schizophrenia). Dementia praecox was normally diagnosed in European adolescents. The theory was that before puberty the cells in the brain's cortex easily bore the acquisition of knowledge. With puberty came the need to exercise the higher processes of the mind, which really tested the durability of the cells. If there was insufficient durability the pressure of the increased demands led to mental collapse; it could occur even in a schoolchild who had previously performed well.[109] Gordon reported that it had traditionally been thought that dementia praecox did not occur in the African, but that while working at Mathari Mental Hospital he had discovered a series of cases in African patients. The cases only occurred in Africans who had been exposed to European education and religion, never in a 'raw native'.[110]

Gordon substantiated this diagnosis of dementia praecox by looking at the brain of a patient who had died; apparently the brain had the microscopic appearances of dementia praecox described by the British expert on the subject, Sir Frederick Mott. Although this was only one sample, it was taken to imply that the African brain lacked the cellular durability necessary to be able to cope with European standards of education. Gordon used Vint's work as well as his own to validate this notion of cellular deficiency.[111] Gordon claimed that the theory that the underdevelopment of the prefrontal cortex left Africans liable to adolescent breakdown under the 'impact of civilisation' was also borne out by his patients at Mathari. Gordon found that of his African adolescent patients at Mathari about whom reliable information could be obtained, all were educated.[112]

The medical leaders and eugenic discourse in the East African Medical Journal

Although the work of Gordon and Vint was the most publicised, they were not the only members of Kenya's medical community who showed an interest in eugenics and racial theories. Doctors of high professional status actively supported the research into mental backwardness and showed eugenic sympathies. Two directors of Medical and Sanitary Services in Kenya (Dr Gilks and Dr Paterson), two directors of Medical Services in Uganda (Dr Kauntze and Dr Trowell) and the long-serving editor of the *EAMJ* (Dr Sequeira) were involved. Eugenic thinking was therefore more than peripheral among Kenyan doctors; it was expounded by the most powerful people within the profession and engaged them enough to make consistent efforts on behalf of Gordon and Vint's research.

Dr Gilks was Director of Medical and Sanitary Services until his retirement in 1933, when he returned to England and was elected a member of the Council for the BMA representing the local BMA branches of fourteen countries, including the East African territories.[113] In 1936 Gilks became Deputy Chairman of the Dominions Committee of the BMA, and in 1934 he also joined the Council of the British Eugenics Society.[114] Gilks became an important part of Gordon's campaign because of his seniority and because he provided a base in Britain. Gilks was also well known for the work that he did with Dr Orr on nutrition in Kenya.[115] Such nutritional work reflected Gilks' interest in public health and its wider social implications. This was expressed in an article he wrote for the *Journal of the African Society* in which he argued that poverty was neglected when considering public health in rural Africa and that economic development would be crucial in bringing about substantial improvements.[116]

Gilks was central to the mental capacity campaign in Britain. He began the pursuit for government funding and support with his proposal, submitted in June 1931, for the continuation of research into African backwardness by means of a grant from the Colonial Development Fund. He wrote another, slightly less ambitious proposal along similar lines in 1932, following Dr Sequeira's success in arousing the interest of various

British eugenicists and doctors.[117] Gilks' enthusiasm, combined with his influential position as retired Director of Medical Services, was vital in gaining attention for the debate on race and mental backwardness. He continued to be active in the public debate after his retirement in Britain, as will be shown in the next chapter.

Another important medical figure who supported Gordon and became involved in eugenics in Kenya was Dr James Sequeira, FRCP, FRCS. In his position as editor of the *EAMJ* Sequeira demonstrated his support for Gordon in editorials. He also became Vice-Chairman of the KSSRI when it was formed in 1933.[118] At the founding meeting of the Society, Sequeira made a speech on the importance of eugenics, arguing:

> that Man should take a hand in his own evolution to improve his own race as he improves that of horses, dogs and cattle ... All men are not equal, either by nature or nurture ...[119]

The main concern that Sequeira ventilated at KSSRI meetings was about future generations of European children. He often warned of the development of a poor white class in Kenya and argued that the way to prevent this development was by 'diminishing the numbers of the poorer part'. Kenya needed to develop a 'eugenic conscience' to solve these problems.[120]

Sequeira had had a distinguished career in Britain until he moved to Kenya because of ill health on his retirement in 1927. He had been Physician to the Skin Department at the London Hospital, and when he left Lord Knutsford addressed the House of Commons in a tribute to his achievements. In this address Knutsford congratulated Sequeira on having introduced X-ray treatment and phototherapy to the London Hospital; he was a member of the Council of the College of Physicians and was President of the Dermatological Section of the Royal Society of Medicine; for years he was editor of the *British Journal of Dermatology* and his *Diseases of the Skin* was a classic textbook.[121] In 1946 he was elected an Honorary Member of the Royal Society of Medicine.[122] The Kenyan BMA was obviously impressed and gratified by the presence of such an eminent doctor in their circle, as reflected in the Governor's opening speech at the centenary meeting of the Kenyan branch of the BMA in 1932:

> There could be no better advertisement for the Highlands of East Africa than the fact that Dr Sequeira and other eminent medical men have on retirement decided to come and live among us; and I am personally aware of the self-sacrificing manner in which they all so freely place their special knowledge at our disposal.[123]

A statement of Sequeira's eugenic sympathies can be seen in his article 'Heredity and Disease: The Story of the Abbé Gillet's Rabbits'. This article is an exposition of Mendelian genetics and, with its hint of Lamarckism, was of the kind espoused by what Kevles describes as the 'mainline eugenicists' and increasingly dismissed by geneticists as their understanding of heredity developed.[124] Sequeira's argument in this article was that dystrophy caused by infection could become hereditary and thus have effects on future populations:

> I conclude with the question that if a dystrophy due to one form of infection can follow the Mendelian laws of heredity is it not possible that some other types of microbic invasion may also act as race-poisons?[125]

Sequeira held the traditional and slightly out-of-date (in terms of contemporary genetic knowledge) ideas that typified British 'mainline eugenics' in the 1930s. Sequeira had connections with some distinguished British eugenicists and doctors; he wrote a letter of introduction for Gordon to Sir Humphrey Rolleston (President of the Eugenics Society).[126] Sequeira was also a friend of Lord Dawson of Penn, Professor Arthur Keith, and Professor Elliot Smith, who were all extremely eminent British doctors and anatomists, as well as R. Langdon-Down (a Vice-President of the Eugenics Society) and Alfred Tredgold.[127]

Sequeira therefore provided important and prestigious links between Kenyan and British eugenics. In the spring of 1932, he spent several months in Britain, during which time he attempted to arouse interest in the research on the East African native: he wrote a short article in the *BMJ* on intelligence and race and delivered the 'Chadwick Public Lecture'. This article was an important introduction of the Kenyan thinking on brain and race to the metropolitan audience[128] and provoked a spate of responses in British newspapers, ranging from the *Observer* to local press like the *Yorkshire Post* and the *Glasgow Herald*.[129]

Through the campaign for funding of research on race and intelligence we find evidence of the eugenic thinking of other East African doctors who otherwise would appear to have been uninterested in such questions. An example of this is Dr Kauntze, who was made Director of Medical and Sanitary Services in Uganda in 1933 and was on the editorial committee of the *EAMJ*. Kauntze went on to become Chief Medical Adviser to the Colonial Office in London. He was 'a doctor of strong humanitarian principles' and actively encouraged the training of African laboratory assistants.[130] In the report on mental deficiency at Kabete Reformatory, Gordon acknowledged the help given by Dr Kauntze in discussing difficulties and giving advice.[131] Kauntze was considered a part of the team who would be involved in the prospective research campaign on the brain of the East African native, and had been consulted on the plans for the project, which Gilks sent to the Colonial Office.[132]

Kauntze also seems to have been important in the initiation of Vint's research on the prefrontal cortex of the East African native, as Gordon put it, 'Dr Kauntze, scientist and prince of silent organisers, had instituted this research by Dr Vint'.[133] Kauntze did not himself contribute any articles on eugenics or race and intelligence to the *EAMJ* and it would be presumptive to label him a eugenicist. There is evidence, however, that he had firm confidence in the importance of continuing Vint's research, as seen in his comments at the discussion that followed Vint's presentation of his paper on the cell content of the prefrontal cortex:

> Dr KAUNTZE congratulated Dr Vint on his paper for its lucidity, its restraint in drawing conclusions, and its absence of bias … He pointed out that the solution of these problems involved a vast amount of work, and suggested that the subject was of such importance as to merit a grant from such a fund as the Carnegie Trust which would allow of the employment of a pathologist to relieve Dr Vint of routine duties, and enable him to devote his whole time to further investigation of the African brain.[134]

At the Kampala meeting of the East African branches of the BMA in 1936, a resolution was unanimously passed which stated that the solution of the question of African capacity was a matter of urgency and that research along the lines recommended by Gordon should be conducted. The resolution was proposed by Sir Albert Cook, President of the Conference, and seconded by Kauntze.

The involvement of Kauntze brings us to a wider problem in defining eugenic affiliation in Kenya. Although there was a high correlation between an interest in this research and in support for eugenics, as can be seen in the cases of Gordon, Gilks and Sequeira, it is difficult to pinpoint where an association with eugenic concerns and being a committed eugenicist diverge. This is particularly true in the case of racial issues: the language of biology and the attribution of cultural and social problems to biological causes were widespread in Kenya's European population. Interest in racial questions was such that eugenics, and more generally the language and *mentalité* of eugenics, seem much more diffuse in Kenya because of the extent of the cultural consensus on race. Race was such a preoccupation in colonial Kenya, and the biological attitude towards race so rooted, that eugenic language seems omnipresent, even among people who would not necessarily have described themselves as eugenicists.

The case of Dr Paterson provides an example of the difficulties in defining commitment to the eugenic idea. Paterson was Deputy Director of Medical and Sanitary Services in Kenya from 1920 to 1933, when he succeeded Dr Gilks as the Director, a post he held until his retirement in 1943. Paterson was embroiled in the campaign for the funding of a research project on the East African brain, and yet, despite this involvement in eugenics in Kenya, much of his writing tended to environmentalism in its approach to the causes of racial differences. This can be seen in his response to Vint's work during the discussion at the end of the presentation on the cell content of the prefrontal cortex of the East African: 'Were the African living under more hygienic conditions might not the average weight and size be bigger?'[135] Paterson went on to emphasise the extra importance of the provision of adequate education facilities in the light of Vint's findings.[136] Paterson's approach to the problem of so-called backwardness in the East African was very different from Gordon's:

> The term 'backwardness' is used as indicating that the people are for the most part still illiterate, that their methods of agriculture and animal husbandry are primitive, and their general standards of culture and of living low to a degree when compared with modern standards in western Europe; the term is not, however, intended to indicate that the peoples are uncivilized for they have in all cases achieved civilization which is in many ways complex; but their civilization is different from that of Europe and in many respects, though not in all, it is ill-suited to meet the needs of the people under the changing conditions which are arising as a result of new developments and from contact with western peoples and methods.[137]

For Paterson the function of education in Kenya and the purpose of the proposed research lay in aiding the project of African modernisation and development. On the subject of intelligence and development in his 'Memorandum on Development in Kenya' Paterson wrote: 'There is no reason to suppose that African [mental] capacity is unduly low'[138] and went on to say:

> Now we know that the moral and intellectual achievement of a small number of Africans has been shewn [sic] to be very high, and in some regards much higher at least in intellectual achievement than the achievement of the average European ... But with regard to the average African what do we know? We know undoubtedly that the intellectual capacity of a very large number of Africans is very much greater than we had imagined but a few years ago. With a minimum of training, and in spite of insuperable disadvantages in upbringing, thousands of Africans in Kenya today are taking part in the life of the Colony in a fashion which but a few years ago would have been difficult to imagine.[139]

Paterson's approach and its awkward relationship with the 'mainline' Kenyan eugenics is best understood as falling into the social hygiene category of thinking – it encompassed both environmentalist and eugenic attitudes in the pursuit of a biologically improved society.

The example of Paterson indicates the lack of homogeneity in eugenic thinking; his outlook was quite different from Gordon's and yet he was on the committee of the KSSRI.[140] And his environmentalist public health agenda did not necessarily preclude a belief in the reality of mental differences:

> In Africa we are dealing with a people whose 'mental make up', whatever meaning we may give to that term, may be and most probably is, radically different from, though not necessarily inferior to, that of Europeans, and whose environment and traditions are undoubtedly fundamentally different. European methods of education and guidance which are still admittedly gravely defective as regards Europeans may be still more defective if applied to Africans ...[141]

Although Paterson did not fall into the same category as the likes of Gilks and Gordon in his beliefs about African mental backwardness, he was a passionate supporter of the campaign for research funding. As Director of Medical Services he won the support of the Kenya government for the campaign; he also unsuccessfully attempted to acquire the support of the British government. Paterson believed there would be far-reaching social value in continuing the research from a local and global point of view:

> There are three million folk in Kenya, over a hundred million in Africa, and many more millions of similarly 'backward' folk elsewhere, all of whom are now either in contact; or on the brink of contact, with a civilization which for good or ill must have profound effects on their domestic conditions, and their future as individuals or as races ...[142]

Part of Paterson's motivation for involvement in the eugenic research campaign lay in his ambition to establish Kenya as an international centre for medical and scientific research. He was keen on attracting staff of the highest standards to his administration. Stanton, Chief Medical Adviser to the Colonial Office, described Paterson as having a passion for 'celebrity collecting', and the Colonial Office clearly considered it unrealistic to attract such 'supermen' to take up jobs in Kenya.[143] Despite this scepticism within the Colonial Office, medical research, partly thanks to the international publicity that Gordon and Vint's work attracted, did gain some of the reputation sought by Paterson. T. Walter Wallbank, Fellow of the American Social Science Research Council, visited Kenya and wrote a report on his findings in 1937:

> Of all British African colonies, Kenya has been pre-eminent in its efforts to work out a scientific technique for investigating the mental and physical make-up of the African. This may result in a valuable contribution to the

study of racial differences and throw light on what, if any, basic differentia exist between the African and European ... Medical activities among the Africans in Kenya are dominated by the view that work should not be confined to pills and liquid potions but of equivalent importance is the responsibility of raising the general standard of living of the people.

The driving force behind this objective is the Director, Dr A. R. Paterson. I was greatly impressed by his ability to envisage the essential unity of the problem of native progress.[144]

Paterson's approach to the Kenyan research on race and intelligence was to attempt to soften and integrate it into a wider and more progressive social reform agenda; he is a good example, in the colonial environment, of how eugenics could attract liberal reformers as well as social reactionaries.

Paterson's enthusiasm for science was combined with a faith in progress and development, which can be seen in his promotion of public health education as a way of combating backwardness. In 1934 he produced *The Book of Civilization*, a pamphlet that was designed to be distributed to the African population in Kenya to inform them about the importance of matters like hygiene, a balanced diet and the dangers of overcrowding. It also attempted to explain economic matters, like the need to grow surplus crops for markets and the point of taxation.[145] Paterson's ambitious enthusiasm for raising African living conditions through education and his public health schemes do not seem to have been particularly respected by his subordinates in the medical department: his attempt to eradicate the rat population ended up in him becoming the butt of a practical joke by a colleague.[146] The lack of respect towards Paterson seems to have been partly caused by his failure to conform with settler culture: he seems to have been viewed with suspicion, according to the memoirs of Dr Clearkin, because he did not play sports or know much about horses.[147]

There may be a further dimension to this; Paterson's enthusiasm for improving conditions for Africans was on a broad social scale. He envisaged a wider scope for the medical administration and a more comprehensive sense of responsibility for the welfare of the African community in Kenya. Dr Trowell's reminiscences about Kenya in the 1930s indicate that there was a clear distinction among those involved in the administration, or 'native welfare' more widely, between those who were in favour of 'African development' and those who were against it, or at least indifferent. In a recorded conversation Dr Trowell and his wife (Kathleen, KMT) spoke of Paterson in the following manner:

He was the only doctor, the only doctor in the Medical Service, who was really terribly keen on the African. He was frightfully – people said unreal-

istically – idealistic about their potential. All the others regarded him as unrealistic. They were just out there for their twenty years ...
    KMT: 'Just looking after these nigs', they'd say.[148]

Dr Trowell is also another interestingly ambiguous figure; he was given the job of training African hospital assistants in Nairobi in 1932, and in his *Memoirs* discussed the resistance to the medical training of Africans. He described the theories of Gordon and Vint as a 'sinister attack on any plans to educate Africans'.[149] Yet Trowell himself was present at various race improvement meetings, publicly supporting the campaign for more research on intelligence, which is difficult to reconcile with the stance taken in his memoirs.[150] It is clear, however, that Trowell was among the more progressive members of the colonial administration, who along with individuals like Paterson were pushing for native development and saw themselves as being supportive of native interests.

The involvement of individuals such as Paterson and Trowell with the KSSRI was clearly felt to be consonant with their values when it came to native affairs: the relationship between Kenyan eugenics and politics was more complex than Trowell's retrospective interpretation suggests. After World War Two, the issue of race became polarised in a way that it simply never had been in the inter-war period. This is not to argue that the race improvers were not racist, but merely to assert that the scientific racism was still so dominant, particularly in Kenya, that it was somehow unobservable, making it possible for individuals to countenance Gordon and Vint's theories on race and intelligence while still being perceived, and perceiving themselves, as extremely forward-thinking on race. Tom Askwith, a notably progressive and forward-thinking official, who became Commissioner for Community Development in the 1950s, attempted to convey how racism became so pervasive in the East African colonial environment, explaining it on the basis of cultural differences and different medical standards and anxieties:

> Racial discrimination, it was often argued, was not an ethnic but a culture bar, the real reason being the differences in health standards as between Europeans and other races. In order to preserve our health and maintain our standards of living we must keep aloof from those whose way of life was primitive. In an insidious way this mental attitude became acceptable, and I am afraid that in a comparatively short time one found oneself looking on Africans and Indians as inferior people. We were not really surprised when certain scientists propounded the theory that black people were actually intellectually inferior to white ones. We also tended to feel superior to Asians whom we came to regard as deceitful and dishonest. So perhaps the outside world was right in what it said about our racial attitudes, but less correct in the reasons they attributed to their cause. It was not simply that we personally wished to maintain our position of authority as is sometimes

alleged, but mainly that we were genuinely worried about all the diseases we might catch if we mixed with people who had dirty habits.[151]

The cases of Dr Paterson and other progressive doctors like Trowell are significant; they indicate that the theories propounded by the likes of Gordon, Vint, Gilks and Sequeira were more than a scientific expression of belligerent settler racism, although their ideas were certainly associated with that perspective. Eugenics and its application to race and intelligence took root in the Kenyan medical profession partly because it promised rational, biological solutions to perceived social problems, in particular African backwardness and the shape of future African development.

The researches of Dr Gordon and Dr Vint stimulated more reactions than any other question in the *EAMJ* at that time. So far, the influential support for Gordon and Vint's research has been demonstrated. One factor in the lack of extensive criticism of the work of Gordon and Vint may be that the editorship of the *EAMJ* was in the hands of Gilks and then Sequeira. The assistant editor was Dr Carman, who was also a eugenicist and, as will be discussed in Chapter 6, believed in African intellectual inferiority. As assistant editor, Carman selected and edited the proofs, while the writing of the editorials remained in the hands of Gilks or Sequeira; several editorials were dedicated to the work on African intelligence and its importance.[152] The *EAMJ* was the forum for medical debate in Kenya and its domination by doctors whose position on the question of race and intelligence was so strong was likely to affect the discussion within its pages.

There was, however, some significant dissent from Gordon's line, the most important of which came from H. S. Scott, Director of Education in Kenya. Scott had an article published in the *EAMJ* in response to Vint's paper of 1932 on the cell content of the prefrontal cortex of the East African native. Entitled 'A Note on the Educable Capacity of the African', it was a paper Scott also read before the Kenya BMA in May 1932. Scott had previously been Director of Education in the Transvaal and on his retirement from service in Kenya in 1935, he returned to Britain, where he became a member of the Colonial Office Advisory Committee on Education in British Tropical Africa. Scott first of all questioned Vint's correlation of brain weight with intelligence: he argued that different parts of the brain might grow differently, causing amentia and yet not necessarily leading to a lesser brain weight. Scott also attacked Vint's failure to relate the size or weight of the brains he measured to general body measurements. He further pointed out Vint's failure to take into account tribal differences, suggesting that the differences between a Nandi and a Maasai brain may be as great as those between a Kikuyu and a European. Scott argued that the smaller measurements for the brain that Vint had recorded could be attributed to environmental factors, in particular to diet and its effect on growth. He also maintained that education might change such differences, arguing that there was some evidence that the European brain had grown bigger than it had been as little as 500 to 1,000 years ago.[153]

Scott then went on to question Vint's argument about the quality rather than the size of the brain. He argued that Vint's assertion that the East African had obtained approximately 84 per cent of the development of the European, or in other words, that an adult African had the mental capacity of an eight-year-old European boy, should not be seen as terribly shocking considering environmental hindrances, in particular diseases such as malaria and parasitic infections. Scott also pointed out that Vint was working with the statistical averages of a small sample, which did not emphasise the fact that some Africans had greater mental capacity than Europeans.

Scott's final argument was that post-mortem examinations could only provide limited information on intelligence; that the crucial point was to be able to differentiate high from low intelligence in the living. Here Scott refers to the work of Mr Oliver, who had been sent to Kenya by the Carnegie Corporation to devise a standardised intelligence testing system for East Africans. Oliver created a new intelligence test, which was supposed to be a fair test for Africans, and could also be used on Europeans. Oliver's work is described more fully in Chapter 6, but the result of his testing quoted by Scott was that the average score obtained by a European schoolboy was 312, compared to an average score of 266 for the African schoolboys. In percentage terms, the African average was 85 per cent of the European average, which was very close to Vint's figure of 84 per cent when comparing the African with the European histologically. Thus we find that despite his arguments with Vint's methodology and arguments, the premise that the average African was at some level mentally backward was not denied. Where Scott differed was in his interpretation of the findings; he argued that although the average African score might be lower, 14 per cent of Africans equalled or exceeded the European average, and so:

> the evidence as far as it goes does seem to indicate the existence of a large number of Africans who, given good environmental conditions and fair training, are of relatively high educable capacity.
>
> What is the method of approach? I am not concerned with the method of approach elsewhere, but only with our particular method of approach in Kenya. The method of approach may be summed up in the word 'environmentalism'. We are convinced environmentalists.[154]

Scott concluded that it was important that Vint should continue with his work so that more extensive and informative results could be achieved. To do this, he argued that tribal differences be accounted for and that Vint's enquiry be enlarged to encompass adults and children. Scott went on to argue that Vint's post-mortem work should be done alongside intelligence testing, so the findings from the living and the dead could be used to check or confirm one another.[155] Despite Scott's criticisms of Vint's work, he believed it to be significant and valuable, and accepted the main problem of the debate, which was deemed to be the mental backwardness of the African native. Despite his different position on the role of the environment, Scott did not disagree with the framework of the problem, and confirmed that the African brain was a subject of important inquiry. In 1934 he wrote in support of the research proposals to the Colonial Secretary. He went on to say:

> My arguments as against the conclusions of Dr Gordon are that they are based on averages and we have to find out to what extent there is an overlap. If for every 100 Europeans who are fit to be sent to Oxford or Cambridge there are 5 natives, then we shall still need our Makerere. Booker Washington and Aggrey are <u>facts</u>.[156]

Scott is another example of one the more ambivalent supporters of Kenyan eugenics. He supported the research, but was not motivated by the supremacist settler interests; indeed he wrote to Margery Perham of his disapproval of white settlement in Kenya and thought it should not have been allowed.[157]

Scott's environmentalist stance attracted considerable debate and some acerbic disapproval in the pages of the *EAMJ*. Sequeira was the first to jump to Vint's defence in a letter to the journal. Sequeira asserted that although one cannot rely entirely on brain weight or cranial capacity as an index of mentality or intelligence, science did show that:

> more than two-thirds of the mental defectives in white races are small-brained ... When I was in England I showed photographs of Dr Vint's sections demonstrating the paucity of these neurons to various experts and they at once recognized the condition as characteristic of 'high grade' mental defectives.[158]

Sequeira went on to attack Scott's environmentalist approach. Sequeira's line of argument at this point is interesting because it reveals the link between British eugenic treatment of class and Kenyan scientific racism. Essential to both was the idea of nature, of the innate and unalterable flaws of the irredeemably poor and colonised:

> I presume that the 'environmentalist' relies on nurture rather than on nature. In England he would assume that the slum produces the social

inefficient, the 'submerged tenth'. Many years of work in London convinced me that another point of view is tenable, viz., that the mentally subnormal, Karl Pearson's social inefficients, make the slum. Just as I believe the Oriental makes the 'bazaar' conditions and not vice versa.[158]

One of the points of Scott's argument that aroused particular hostility from his medical audience was his emphasis that a proportion of Africans (according to Scott's calculations, 80,000 in Kenya) were equal or superior to the average European in intelligence. Dr Arnell wrote to the *EAMJ* to criticise this statement by Scott on statistical grounds, arguing that Scott was using too small a sample (93 African and 124 European schoolboys were given the intelligence test Scott refers to) – interestingly Vint's sample of 35 for examination of the cell content did not earn the same opprobrium.[160] The last words in response to Scott in the correspondence to the *EAMJ* were given to Dr Gordon. Gordon considered Scott's environmentalist approach antiquated and impractical: 'This is the old view and its practical application for many generations by industrious believers has not cured African backwardness.'[161]

The responses to Scott's riposte to Vint show that Kenyan doctors did not just argue that racial backwardness existed; professional hostility was also provoked by asserting that this backwardness was not caused by heredity. The passionate rebuttal of Scott's criticisms, despite the fact that Scott was keen to show that he basically supported the pursuit of more research, indicates a strong intolerance of deviation from the hereditary interpretation of racial difference. This is not to argue that there was unanimous agreement on Gordon and Vint's researches among doctors in Kenya; missionary doctors, for example, are notably absent in this discourse. However, it is striking that in the 1930s many of the most powerful doctors in Kenya – whose influence derived from high official status (Gilks, Paterson, Kauntze), professional eminence (Sequeira, Vint, Gordon) and control of the *EAMJ* (Sequeira, Carman) – formed a formidable coalition in support of the research on race and intelligence.

Notes

1 'Kenya's Progress in Public Health', *EAS*, 31 January 1931, p. 28.
2 Ibid.
3 G. R. Searle, 'Eugenics and Class' in C. Webster (ed.), *Biology, Medicine and Society 1840–1940* (Cambridge, 1981).
4 Medical Service List, *EAMJ*, 16/8 (1939), pp. 313–19.
5 J. A. Carman, *A Medical History of the Colony and Protectorate of Kenya: A Personal Memoir* (London, 1976), p. 40.
6 H. L. Gordon, 'Amentia in the East African', *Eugenics Review*, 25/4 (1934), p. 235.
7 Petition by Kenyan doctors, 29 September 1936, NA, CO 822/74/10.
8 J. McCulloch, *Colonial Psychiatry and 'the African Mind'* (Cambridge, 1995), pp. 46–9,

and M. Vaughan, *Curing Their Ills – Colonial Power and African Illness* (Stanford, 1991), p. 110.

9 McCulloch, *Colonial Psychiatry*, p. 49

10 S. Dubow, *Scientific Racism in Modern South Africa* (Cambridge, 1995), pp. 120–65.

11 H. L. Gordon, 'The Mental Capacity of the African', *Journal of the African Society*, 33/132 (1934), pp. 27–50, 27.

12 'Henry Laing Gordon, M. D. Edin.', *EAMJ*, 24/9 (1947), pp. 313–14.

13 Interview no. 4, Laurie Slade, 27 January 1999.

14 KNA, PC/Nyanza/1/3/27/17.

15 Ibid.

16 Gordon to Blacker, 17 August 1930, CMAC, SA/EUG/C.129.

17 Mathari Mental Hospital was founded in 1910 on the outskirts of Nairobi. It originally had eight beds for Africans and two for Europeans, although it was never considered really suitable for European patients. In 1930, there were 144 admissions to Mathari. See McCulloch, *Colonial Psychiatry*, pp. 20–8.

18 Annual Medical Report, 1929, p. 34, and Annual Medical Report, 1931, p. 43.

19 Memorandum by Flood, 9 November 1936, NA, CO 822/72/8.

20 'Annual General Meeting of the Kenya Branch of the BMA', *KEAMJ*, 7/11 (1931), p. 312.

21 H. L. Gordon, 'A Note on the Diagnosis of Amentia (Mental Deficiency) in Africans', *KEAMJ*, 7/8 (1930), pp. 208–14.

22 *EAMJ*, vols 5–22 (1928–45).

23 H. L. Gordon, 'The Doctor in the Witness-Box', *EAMJ*, 18/8 (1941), pp. 236–9; 'Is War Eugenic or Dysgenic?', *EAMJ*, 19/2 (1942), pp. 86–96; and 'The Importance of Social Medicine to Kenya', *EAMJ*, 23/1 (1946), pp. 2–12. Also, H. L. Gordon, 'Population Problems in a Crown Colony', 24 May 1945, NA, CO 533/537/13.

24 See obituaries, 'Dr H. L. Gordon', *Eugenics Review*, 39/3 (1947), p. 108, and 'Henry Laing Gordon, M. D. Edin.', *EAMJ*, 24/9 (1947), pp. 313–14.

25 Gordon to Blacker, 17 August 1930, CMAC, SA/EUG/C.129.

26 Gordon to Blacker, 5 October 30, CMAC, SA/EUG/C.129.

27 Gordon, 'Amentia in the East African', p. 235.

28 'The Mind of the Native', *EAS*, 7 October 1926, p. 7.

29 Ibid.

30 Gordon to Blacker, 5 October 1930, CMAC, SA/EUG/C.129

31 H. L. Gordon, 'The Social Aspect of the Curious History of Syphilis', *KEAMJ*, 7/11, (1931), pp. 320–32. See also Gordon, 'Mental Instability among Europeans in Kenya', *KEAMJ*, 4/10 (1928), pp. 316–24, and 'Relation of Malaria to the Alleged Rarity of Neurosyphilis amongst "Uncivilised" Races', *KEAMJ*, 6/8 (1929), pp. 221–9.

32 Gordon, 'The Social Aspect of the Curious History of Syphilis', pp. 326–7

33 Gordon, 'Mental Instability among Europeans in Kenya', p. 322.

34 H. L. Gordon, 'Eugenics and the Truth about Ourselves in Kenya', NA, CO 822/55/1.

35 'To Study Improvement of the Race', *EAS*, 3 June 1933, p. 11.

36 Gordon, 'Eugenics and the Truth about Ourselves in Kenya'.

37 Ibid., p. 2.

38 Ibid., p. 3.

39 Ibid., p. 6.

40 The Reformatory was the only institution exclusively for juvenile offenders in the colony at that time. It was established in 1909 in Kabete, a few miles outside Nairobi. For more on the Reformatory, see Chapter 6.

41 H. L. Gordon, 'Report of a Survey of the Inmates of Kabete Reformatory', CMAC SA/EUG/C.129. Also in KNA, AP/1/701.

42 Gilks to the Colonial Secretary, 3 March 1931, KNA, AP/1/699; Minutes of the Meeting of the Committee of Visitors, 3 December 1930, KNA, AP/1/700.

43 Gordon, 'Report of a Survey of the Inmates of Kabete Reformatory', Table III, p. 13.

44 Ibid., p. 20.

45 Ibid., p. 8.

46 R. J. A. Berry, 'Mental Deficiency in England: An Analysis of the Mental, Physical and Mental Characteristics of a group of 162 Adult Feeble-Minded Women' in R. J. A. Berry (ed.), *Stoke Park Monographs on Mental Deficiency and Other Problems of the Human Brain and Mind* (London, 1933), pp. 65–80, 80.
47 Berry (ed.), *Stoke Park Monographs on Mental Deficiency*, p. vii.
48 Berry, 'Mental Deficiency in England', p. 80.
49 Ibid., p. 67.
50 'Nakuru Murder Trial', *EAS*, 3 December 1932, pp. 44–6.
51 H. L. Gordon, 'The Case of Charles William Ross, Hanged for Murder in Kenya Colony', in Berry (ed.), *Stoke Park Monographs on Mental Deficiency*, p. 237.
52 P. J. Shaw Bolton, *The Brain in Health and Disease* (London, 1914), p. 161, Bolton's italics.
53 Gordon, 'A Note on the Diagnosis of Amentia', p. 208.
54 A. F. Tredgold, *Mental Deficiency (Amentia)* (London, 1929), fifth edition.
55 Gordon, 'A Note on the Diagnosis of Amentia', p. 208.
56 R. J. A. Berry, 'Practical Method for the Detection, During Childhood, of Potential Social Inefficiency and High-Grade Mental Deficiency' in Berry (ed.) *Stoke Park Monographs on Mental Deficiency*, pp. 10, 18.
57 Gordon, 'A Note on the Diagnosis of Amentia', p. 210.
58 Gordon, 'Report of a Survey of the Inmates of Kabete Reformatory', p. 5.
59 Gordon, 'A Note on the Diagnosis of Amentia', p. 211.
60 Ibid.
61 'Discussion' at the end of F. W. Vint's 'A Preliminary Note on the Cell Content of the Prefrontal Cortex of the East African Native', *EAMJ*, 9/2 (1932), pp. 30–55, 52–3.
62 Sequeira to Lord Dawson of Penn, 11 October 1936, NA, CO 822/74/10.
63 BMA notes, *EAMJ*, 18/4 (1941), p. 62.
64 Callanan to Attorney General, 11 December 1936, KNA, BY/15/69.
65 Vint, 'A Preliminary Note' and 'The Brain of the Kenya Native', *Journal of Anatomy*, 68/2 (1934), pp. 216–23.
66 *EAMJ*, vols 5–22 (1928–45).
67 F. W. Vint, 'Gall Stones in a Native Child', *KEAMJ*, 5/2 (1928) pp. 59–61; 'The Measurement of Red Blood Corpuscles', *EAMJ*, 16/8 (1939), pp. 295–307; and 'Solar Rays. Fact or Fiction?', *EAMJ*, 21/8 (1944), pp. 227–39.
68 Vint, 'The Measurement of Red Blood Corpuscles', p. 307.
69 F. W. Vint, 'Cirrhosis of the Liver of the East African Native', *EAMJ*, 7/12 (1931), pp. 349–75.
70 *EAMJ* (1932–33), volume 9.
71 This combination of apparent medical neutrality and inevitable racial categorisation fits into a wider debate about the contradictions of colonial medicine; see D. Engels and S. Marks (eds), *Contesting Colonial Hegemony: State and Society in Africa and India* (London, 1994) and Vaughan, *Curing Their Ills*.
72 Comment by Flood, 17 May 1937, NA, CO 822/78/19.
73 R. Crewdson Benington, (prepared for press by K. Pearson), 'A Study of the Negro Skull with Special Reference to the Congo and Gaboon Crania', *Biometrika*, VIII, parts III and IV (1912), pp. 292–337.
74 Vint, 'A Preliminary Note', p. 36.
75 Ibid., p. 48.
76 Ibid., p. 37.
77 Constanin Von Economo was Professor of Neurology and Psychology at the University of Vienna. His texts published in English included *The Cyroarchitectonics of the Human Cerebral Cortex* (London, 1929).
78 Vint, 'A Preliminary Note', p. 46.
79 Ibid.: compare Table I, p. 44, with the later comment, 'Bolton's figure for a normal European is .83 mm', p. 46.
80 Ibid., p. 46.
81 Ibid., p. 35.
82 Ibid., p. 49.

83 Ibid., p. 30.
84 Ibid., p. 52.
85 *EAMJ*, 16/4 (1939), p. 190.
86 Vint, 'The Brain of the Kenya Native'.
87 Ibid., p. 216.
88 Ibid.
89 Paterson to the Acting Colonial Secretary, 1 July 1933, KNA, AG/32/231.
90 Gordon, 'Amentia in the East African'.
91 Ibid.
92 Ibid., p. 226.
93 Ibid., p. 227.
94 Ibid., p. 229.
95 Ibid., p. 232.
96 Blacker to Horder, 13 November 1933, CMAC, SA/EUG/C.129.
97 Matheson (secretary of the African Research Survey) to Oldham 14 December 1933, RH, Afr.s.1829/1/3.
98 Gordon, 'Amentia in the East African'.
99 BMA Council, 1931–32, p. 261, BMA archives.
100 'Dismissal of Dr H. L. Gordon', written by his son and sent to the Eugenics Society, 19 April 1937, p. 4, CMAC, SA/EUG/C.130.
101 KNA, AG/24/28.
102 'Report of the Board of Enquiry, Mathari Mental Hospital', p. 3, KNA, Health/2/115.
103 H. L. Gordon, 'The Intentional Improvement of Backward Tribes', *EAMJ*, 11/5 (1934), pp. 143–56.
104 Ibid.
105 Ibid., p. 143.
106 Ibid., p. 146. Gordon's italics.
107 Gordon, 'The Mental Capacity of the African'.
108 D. Killingray and S. Ellis, 'Introduction', *African Affairs*, 99/395 (2000), pp. 177–82, 177.
109 Gordon, 'The Mental Capacity of the African'.
110 Ibid.
111 Ibid., p. 239.
112 H. L. Gordon, 'An Inquiry into the Correlation of Civilization and Mental Disorder in the Kenya Native', *EAMJ*, 12/11 (1936), pp. 327–35, 231.
113 *EAMJ*, 12/1 (1935), p. 31.
114 *EAMJ*, 13/8 (1936), p. 253; Blacker to Gordon, 11 April 1934, CMAC, SA/EUG/C.130.
115 J. L. Gilks and J. B. Orr, 'The Nutritional Condition of the East African Native', *EAMJ*, 4/3 (1927), pp. 85–90, 89.
116 J. L. Gilks, 'The Relation of Economic Development to Public Health in Rural Africa', *Journal of the African Society*, 34/134 (1935), pp. 31–40.
117 Gilks to Stanton, 28 May 1932, KNA, BY/26/7.
118 Enclosed in Letter from Shaw to Blacker, 4 July 1933, CMAC, SA/EUG/C.129.
119 'To Study the Improvement of the Race', *EAS*, 3 June 1933, p. 11.
120 'Study of Race Improvement', *EAS*, 8 July 1933, p. 46.
121 Editorial, *EAMJ*, 19/1 (1942), pp. 1–2.
122 'Dr J. H. Sequeira', *EAMJ*, 24/3 (1947), p. 141.
123 Speech by the Governor, 5 October 1932, KNA, GH/7/30/BMA
124 D. Kevles, *In the Name of Eugenics: Genetics and the Uses of Human Heredity* (Cambridge, 1995), p. 145.
125 J. H. Sequeira, 'Heredity and Disease: The Study of the Abbé Gillet's Rabbits', *EAMJ*, 10/11 (1934), pp. 331–5, 335.
126 Sequeira to Rolleston, 30 August 1933, CMAC, SA/EUG/C.129.
127 Gordon to Blacker, 4 May 1935, CMAC, SA/EUG/C.130
128 J. H. Sequeira, 'The Brain of the East African Native', *BMJ*, 1 (1932), p. 581.
129 'The Brain of the East African Native', *EAS*, 30 April 1932, p. 45.
130 J. Iliffe, *East African Doctors* (Cambridge, 1998), p. 47.

131 Gordon, 'Report of a Survey of the Inmates of Kabete Reformatory', p. 22.
132 Gilks to Stanton, 28 May 1932, NA, CO 822/47/5.
133 'Discussion' at the end of Vint's 'A Preliminary Note', p. 53.
134 Ibid., p. 55
135 Ibid., p. 54.
136 Ibid.
137 A. R. Paterson, 'The Education of Backward Peoples', *EAMJ*, 8/11 (1932), pp. 302–15, 302.
138 Ibid.
139 Ibid.
140 Enclosed notice with letter from Shaw to Blacker, 4 July 1933, CMAC, SA/EUG/C. 129.
141 Memorandum from Paterson to Colonial Secretary, 23 January 1934, NA, CO 822/61/14.
142 Ibid.
143 Notes by Dr T. Stanton, 27/11/34, and C. A. L. Cliffe, 22 March 1934, NA, CO/533/ 444/5.
144 T. Walter Wallbank, 'American Reflection on Kenya' (1937), NA, CO/533/486/9.
145 A. R. Paterson, *The Book of Civilization*, in the possession of Peter Paterson, Nairobi.
146 Clearkin, *Ramblings and Recollections of a Colonial Doctor, 1913–1958*, p. 146, RH, MSS.Brit.Emp.r.4.
147 Ibid., p. 146.
148 Transcript of conversation between Hugh Trowell (and Trowell's wife, Kathleen Trowell) and his daughter, Elizabeth Bray, in 1982, pp. 4–5. RH, MSS.Afr.s.1872.
149 Trowell, *Memoirs*, p. 9, RH, MSS.Afr.s.1872/XXXIV.
150 'To Study Improvement of the Race', *EAS*, 3 June 1933, p. 11.
151 Askwith, *Memoirs*, p. 5, RH, MSS.Afr.s.1770 (2). See Askwith's published material for a fuller sense of the work and outlook of a progressive, liberal colonial official in Kenya, *From Mau Mau to Harambee: Memoirs and Memoranda of Colonial Kenya* (Cambridge, 1995) and *The Story of Kenya's Progress* (Nairobi, 1953).
152 Editorial, *EAMJ*, 9/2 (1932), p. 29; *EAMJ*, 10/12 (1934), p. 349; *EAMJ*, 11/2 (1934), pp. 39–42. For a description of his involvement in the *EAMJ*, see Carman, *A Medical History*, Chapter IX, pp. 91–8.
153 H. S. Scott, 'A Note on the Educable Capacity of the African', *EAMJ*, 9/4 (1932), pp. 99–110.
154 Ibid., p. 108.
155 Ibid., p. 110.
156 Scott to the Colonial Secretary, 3 February 1934, KNA, BY/26/7.
157 Scott to Perham, 23 October 1930, RH, MSS Perham, Box 474, File 1.
158 J. H. Sequeira, 'Correspondence. The Educable Capacity of the African', *EAMJ*, 9/5 (1932), pp. 146–8, 146.
159 Ibid., p. 147.
160 O. R. Arnell, 'Correspondence', *EAMJ*, 9/6 (1932), pp. 179–80, 180.
161 H. L. Gordon, 'Correspondence. The Educable Capacity of the African', *EAMJ*, 9/7 (1932), pp. 210–12, 210.

Metropolitan responses

In the 1930s the Kenyan eugenicists campaigned for funding from Britain to establish a large research programme in Nairobi dedicated to investigating the causes of African mental backwardness. The highest levels of the Kenyan and British governments and some of Britain's most eminent scientists and eugenicists became embroiled in this campaign. The reception of the Kenyan racial theories in Britain is the subject of this chapter, uncovering the political complexity of the question of 'native mentality'. Confusion over the meaning of the research at first succeeded in attracting a powerful amalgam of voices to the campaign. As the debate in Britain unfolded, the contradictions within the campaign over native welfare, development and the eugenic emphasis on innate inferiority became more crystallised, undermining the Kenyan eugenicists' claims of legitimacy in the name of trusteeship.

In November 1933 letters were sent under the auspices of the British Eugenics Society to *The Times*, the Colonial Office and the Economic Advisory Council (EAC) which sparked off debate about the Kenyan research on race and intelligence. (The EAC was a body working under the Privy Seal, sometimes able to produce funds for work of various kinds.) The signatories to the letters to the Colonial Office and the EAC were Lord Dawson of Penn, Lord Horder, Sir Humphrey Rolleston, Sir Arthur Keith, Grafton Elliot Smith, Julian Huxley, F. A. E. Crew and Dr Gilks.[1] Dawson, Horder and Rolleston were very eminent doctors, Horder and Rolleston were both physicians to the royal family. Keith and Elliot Smith were leading anatomists and physical anthropologists. Frank Crew was Professor of Genetics at the Animal Breeding Research Institute at Edinburgh University. He was one of the key reform geneticists who warned of the dangers of underestimating the role of environment in human behaviour and development; Huxley was Professor of Zoology at King's College London and head of the Zoological Society of London. Between them, Huxley and Crew, who were both among the foremost

British experts on genetics and critics of the mainline eugenic tradition, represented a powerful vote of support.

Despite the different approaches to eugenics taken by the signatories, they were all members of the Eugenics Society, Rolleston was at that time its President and Horder went on to become President in the mid-1930s. Huxley was to become President in 1962. The signatories to the letter sent to *The Times* were fewer: Dawson, Horder, Rolleston, Keith and Elliot Smith.[2] The contents of all three letters were similar, although the letter to *The Times* was more temperate:

> The investigations carried out by Dr H. L. Gordon and Dr F. W. Vint into the mind and brain of the East African native … call, in our opinion, for wider publicity and closer study than they have yet received. The results of these investigations point to the urgency for further and fuller inquiry than has yet been possible with the limited resources hitherto available. A successful native policy can only be laid down on a sure foundation of ascertained fact, and a correct evaluation of the capacity of the peoples concerned. This, in our opinion, can be established by a team of expert scientists, following the lead given by Dr H. L. Gordon and F. W. Vint. We therefore commend this matter to the attention of the authorities concerned.[3]

Unlike the letter to *The Times*, those letters addressed to the EAC and the Colonial Office explicitly mentioned African inferiority. Huxley agreed to be a signatory to the letters to the EAC and Colonial Office demanding funding for research into African intelligence on the condition that the wording was moderated. The original copy said that the results of research so far demonstrated 'a definite degree of inferiority in the average native as compared with the European'.[4] Huxley was unhappy with the use of the word 'inferiority' and Gordon and the Eugenics Society agreed to change the wording. Gordon met up with Huxley to discuss the letter and they appear to have got on well.[5] The final letters did retain the statement about inferiority but qualified it with a comment on the uncertainty about the different factors that might cause it:

> The results arrived at so far tend to show the existence of a definite degree of inferiority in the average of at least certain native tribes as compared with the average European. It is, however, quite uncertain how much of the difference is truly genetic, and how much due to environmental conditions, notably nutrition and disease. It is obvious that the issues raised are of enormous importance, and will fundamentally concern the policy which will be laid down for the development of the East African territories, and for promoting the social advancement of the native.[6]

The first section of this chapter will examine how it came about that the leaders of the British Eugenics Society, who were at that time attempting to reform the organisation along more moderate lines, wrote these letters and so actively supported the Kenyan research. The relationship between Kenyan racial theories and British eugenics is analysed and contextualised. The next section will follow the aftermath of the publicity given to the Kenyan racial theories. There was fierce debate in the correspondence pages of *The Times* and other publications in the winter of 1933/34 on the issue of race and intelligence in relation to the Kenyan research. The Kenyan eugenicists sparked a fascinating metropolitan debate on race and science, which eventually served to diminish the support they had initially enjoyed. The final section will discuss the reaction within the Colonial Office and the British government to the Kenyan eugenics movement.

The response of the Eugenics Society

British eugenics provided the framework for the Kenyan theories on race and the Kenyan eugenicists were eager to establish links with the Eugenics Society. The Society was, in turn, clearly attracted by the Kenyan ideas and willing to encourage eugenics in Kenya. Given that the most conspicuous aspect of Kenyan eugenic thinking was its concern with race, the interaction between Gordon, Sequeira and Gilks (the most active propagandists for the Kenyan work on race and intelligence in Britain) and the British eugenics movement has implications for the connection between British eugenics and scientific racism. An ambiguous and complex intellectual relationship developed as the Kenyan theories were discussed and elaborated within the eugenic constituency in Britain; responses to Gordon and Vint's research did not always follow simple ideological faultlines determined by a racist or antiracist stance. Important figures who became associated with the antiracist attack on mainline eugenics, such as Huxley and Crew, up to a point supported the Kenyan research, while Cyril Burt, a traditional, conservative eugenicist, dismissed it immediately. Criticisms of race in biology were rather unevenly unfolding in the 1930s, and the interaction between British science and Kenyan eugenics must be understood historically within this developing discourse on race in science.

To understand the relationship between British and Kenyan theories of race, the context of the position of the Eugenics Society and the status of scientific racism need to be discussed. As described in Chapter 2, the 1930s saw substantial changes in the British eugenics movement. There was an attack from generally, but not exclusively, left-wing scientists who demonstrated the political biases and class prejudices of British eugenics and argued that the role of the environment had been neglected to a misleading degree when explaining human difference and inequality. Biologists, in particular Lancelot Hogben, Julian Huxley and J. B. S. Haldane, started questioning aspects of scientific racism and eugenics from both scientific and political perspectives. The rootedness of race in science, which had taken hold in the late nineteenth century, was now being undermined by some of Britain's leading experts on human biology and heredity. Much of the criticism of mainline eugenics focused on the inheritance of mental traits and conceptions of mental deficiency. Population geneticists now understood that the inheritance of mental qualities did not follow the simple Mendelian pattern that described plant and animal inheritance of traits like height and colour. Heredity was complicated by factors such as dominant or recessive genes, or sex-linked, or polygenic traits.[7] Throughout the 1930s, Penrose's influential Colchester Survey compounded this understanding of the complex, indirect and multifactoral nature of the inheritance of mental characteristics. Although much of this new thinking in Britain was not directly concerned with race, it had immediate implications for the validity of biological theories on race, with their emphasis on the issue of innate, hereditary intelligence. Scientific racism and mainline eugenics thus became connected in the criticism produced by the new genetics of the 1930s.

The Eugenics Society was surprisingly resilient in the face of this growing body of criticism, helped, as Searle has argued, by external factors such as a defensive response to the increased poverty and unemployment caused by the Depression.[8] Partly thanks to Blacker's flexible leadership, the movement regained some academic vigour and credibility through the emergence of what Kevles has termed 'reform' eugenics.[9] This attempted to reconcile many of the experts who criticised the excesses of mainline eugenics and accommodated a more temperate approach based on the new appreciation of the complexity of human heredity.

Reform eugenics and the growing critique of scientific racism did not have entirely negative implications for the reception of the Kenyan theories by British eugenicists, at least in the short-term. Partly because of the longstanding link between eugenics and genetics in Britain, much of the criticism of mainline eugenics came from within the movement, from individuals such as Ronald Fisher and Julian Huxley – even the more radical Haldane retained a place for eugenics in genetic science. The process by which the demise of eugenics and scientific racism occurred was not straightforward and in the debates over race in science in the 1930s, scientific research that purported to address the question objectively was welcomed by many biologists and eugenicists. It was increasingly argued that variation within racial groups could be enormous and that individuals of high intelligence existed within any group, but that there were still differences in the average intelligence of racial or ethnic groups. As Julian Huxley put it: 'the *average* differences which do exist between races are real enough, and have often a definite biological significance.'[10] Huxley's support for Kenyan eugenics was crucial, for he held an important position as an expert on scientific matters in Britain and had influence over the more liberal side of the Eugenics Society. Huxley also had direct connections with East Africa; he spent time there in 1928 collecting information for a memorandum he was asked to write on biological teaching in African schools by the Colonial Office Advisory Committee on Education in Tropical Africa. Huxley visited Tanganyika, Zanzibar, Uganda and Kenya. In Nairobi, he inspected the Medical Research Laboratories, where Vint conducted his research on the African brain.[11] Huxley's other connection to Kenya was that his cousin, Gervais Huxley, married Eleanor Grant's daughter, Elspeth,[12] although it appears that there was little love lost between Julian and the Grant family.[13]

The Kenyan doctors' claim to be engaged in rigorous, objective scientific research into the role of biology in African backwardness was entirely compatible with the programme of reform eugenics. Blacker was eager to bolster the legitimacy of the Eugenics Society by aligning it with serious exploration of matters relevant to social policy. The Kenyan proposals made enthusiastic claims about the social applications of the research, which on the surface appeared to chime with the desire for investigation into African conditions expressed by eminent metropolitan experts on Africa, such as Dr J. H. Oldham, Secretary of both the Conference of British Missionary Societies and the International Missionary Council, and academics in Oxford linked with the *African Survey* project, including Margery Perham and Professor Coupland.

The situation changed later on in the 1930s as the arguments against scientific racism became increasingly articulated, but when Gordon first wrote to Blacker, the approach he presented appeared eminently compatible with the research emphasis of reform eugenics. Gordon entered into correspondence with Blacker in 1930 and they discussed the question of race and intelligence almost immediately. Blacker was relatively cautious on the subject to begin with:

> the admitted lower intelligence ratio of African natives as encountered in the descendants of slaves in the U.S.A. may be due to a fairly wide breeding from aments, and not to any inherent incapacity in intelligence in the Black races e.g. the descendants of the ruling families, some of whom I have met, appear to have a very good intelligence as compared with white civilized races.[14]

Gordon answered that a 'submerged group', in this case aments, was abundantly present in the entire native population of Kenya, but that it was disguised by lack of civilisation and low cultural expectations. Blacker seems to have been convinced by this.[15] Gordon made it clear to Blacker that he was pursuing the question of amentia in Kenya from an objective point of view; he was careful to distance himself from the settler perspective, which had become notorious in Britain in relation to native interests.[16] Gordon even went so far as to comment that on an individual level the correlation between race and amentia was not useful, it was when it came to assessing overall trends and attempting to pursue the white man's burden of raising the overall social standard that it became useful:

> Clinically, race may make no appreciable difference to amentia; socially, amentia makes such difference to a race. It is a public health problem in mental hygiene involving psychiatry, psychology, biology, sociology, eugenics and euthenics; a problem of deep concern to jurisprudence, morality and religion. The higher the type of society the higher the type of social reaction demanded of the individual ...[17]

This approach has some parallels with Julian Huxley's view on race in the early 1930s: that average differences between races could be a biological reality and of social significance, despite the fact that individuals should not be regarded as limited by their race.

At the end of 1930, Gordon sent a copy of his 'Report of a Survey of the Inmates of Kabete Reformatory for the Purpose of Detecting the Presence of Amentia (mental deficiency)' to the British Eugenics Society.[18] This was the study in which Gordon concluded that the vast majority of the inmates of the Reformatory for African boys in Kenya were either high-grade or medium-grade aments, leading him to assert that there was a serious problem of innate mental backwardness in the East African 'native'. The paper went down well with the Eugenics Society in Britain; it was described as being 'of great interest and value'.[19] Gordon continued his correspondence with Blacker and the Eugenics Society gave its support and promised to do its 'utmost to assist' the newly formed KSSRI, offering to send literature and a set of back copies of the *Eugenics Review*.[20] Both Blacker and Gordon were psychiatrists and it is clear that Blacker felt that on a personal and professional level, Gordon was an appropriate person for the Eugenics Society to support; Blacker described Gordon to Julian Huxley as a 'charming man personally, and, in my opinion, entirely safe to back'.[21]

The links between the Kenyan eugenicist doctors and the British Eugenics Society intensified in 1933 when Gordon was given special leave by the Kenyan medical department to visit London to publicise his theories and to pursue the possibility of funding for his research. During his stay in London, Gordon addressed a meeting of the Eugenics Society on 7 November 1933 with the paper 'Amentia in the East African', which was published in the *Eugenics Review* in January 1934.[22] In this paper, Gordon gave his findings on measurements of brain capacity, as well as measurements of brain weight, and head length, breadth and height. Gordon was using methods that had largely been discredited by biologists in Britain, and yet the Eugenics Society did not seem to be put off by this. In fact the paper was a great success, as was conveyed to Lord Horder:

> Gordon gave an excellent address to a Members' Meeting last Tuesday, and he was congratulated on his work by Professor Seligman, Sir Hubert Bond, Dr A. F. Tredgold and some representatives of the Colonial Office. I don't think there is any doubt that he is a safe man to back.[23]

Seligman was Professor of Ethnology at London University; he was interested in race psychology and an influential proponent of the Hamitic idea in African studies.[24] Unfortunately who the representatives of the Colonial Office were, and a clearer idea of what they thought about the paper, is not revealed in the documents on the subject in the Public Record Office or in the Eugenics Society archives.

During the campaign for research into race in Kenya, the Eugenics Society managed to obtain the support of some of the more critical and forward-thinking biologists in Britain, scientists who were to have a role in challenging scientific racism and the more extreme conservatism of eugenic thought. The link that enabled the Kenyan eugenicists to gain the endorsement of Crew seems to have been Fraser Roberts, who wrote to Crew requesting his support.[25] The fact that Gordon had rallied Roberts to his cause is a fascinating indication of the range of allies the Kenyan programme acquired. Roberts became a leader in human genetics in Britain after doing post-graduate studies under Crew in Edinburgh and then working at Stoke Park Colony for the Mentally Defective in Bristol. Roberts' work deliberately dovetailed with Penrose's work in Colchester, he was a typical reform eugenicist, who, like Penrose, did much to work towards clearing misconceptions about heredity and mental deficiency. His book, *An Introduction to Medical Genetics*, first published in 1940, was highly influential and a standard text for many years.[26]

The nature of the support given to the Kenyan eugenicists creates a complicated picture in which Gordon's ideas, at least in the earlier stages of their publicity in Britain, were by no means associated exclusively with the more reactionary eugenicists. This was partly because the eugenics movement in Britain in the early 1930s was in a state of flux. Ideas about human genetics, race and heredity were undergoing serious reconsideration, but in the early 1930s, this discourse was still in its early stages and the criticism had not yet developed to the point of entirely rejecting the old language of race and eugenics. Many scientists who later seriously undermined old conceptions were in the early 1930s still engaging with them, albeit with a critical eye.

Blacker consistently attempted to reconcile the left-wing scientific critics of the movement, inviting Haldane and Hogben to participate in the society, and inviting Penrose to give a lecture. In 1935 Gordon wrote to Blacker asking him to recommend a good geneticist who might help with the research into African backwardness; Blacker suggested Penrose.[27] This choice of geneticist is an intriguing reflection of the current state of the eugenics movement – Penrose's work on the Colchester Survey contributed an important assault on conventional eugenic ideas about inheritance and mental deficiency. As Barkan argues, Penrose also played an important role in the attack on scientific racism, partly through the scientific principles he introduced in his work on mental deficiency, which undermined 'old taxonomies which had no biological basis but social and cultural bigotry'.[28] Penrose also challenged the term 'race' as biologically unsound. The fact that Blacker suggested Penrose as an appropriate geneticist for Gordon's purpose indicates Blacker's faith in the shared pursuit of scientific truth and perhaps a certain naïveté about the level of prejudice within Kenyan eugenics. Unfortunately, there is no record of Penrose's response to Gordon's theories; he was clearly not one of Gordon's forthright supporters.

Julian Huxley's changing stance on race has been described in Chapter 2; his attitude towards mental capacity in Kenya more specifically is interesting. His support for Gordon and Vint's work is expressed in his involvement in the signing of the Eugenics Society's letters of November 1933 to the EAC and the Colonial Office and he confirmed his support for the cause in a further letter to *The Times* defending the Kenyan research in December 1933.[29] In July 1933 Huxley had written the preface to a book by Parmenas Mockerie, which was published in 1934.[30] Mockerie was considered dangerously subversive in Kenya: he was a founder of KISA (the Kikuyu Independent Schools Association), a delegate of the Kikuyu Central Association, and with Kenyatta was sent to London to give evidence before the Parliamentary Committee on East Africa in May 1931. In his foreword, Huxley argued that Mockerie himself was proof of the innate capacity of Africans and drew attention to the injustices felt under 'the best-intentioned acts of paternalistic government',[31] mentioning the *kipande* system (the certification system enforced under the Registration of Natives Ordinance, which required all adult African men to carry a registration pass at all times while outside the reserves), the restrictions on public meetings and political activity, and low wages. Huxley also mentioned in the foreword the belief in African mental inferiority and stated that it was impossible to say whether such a position was correct or not. He went on to say that the African was clearly far more capable of benefiting from education than believers in white superiority had held. The fact that Huxley was able publicly to support Gordon's campaign for research into race and intelligence in East Africa just a few months after writing the preface for Mockerie's book captures the complexity of motivation behind the popularity of the Kenyan eugenicists' thinking on intelligence. Indeed, Huxley was still supporting Gordon in 1938, long after many regarded him to be discredited in Britain and 'Hogben called us charlatans and Malinowski held us up to ridicule'.[32]

British experts on amentia who more closely followed the thinking of mainline eugenics also took up the Kenyan theories, making them a part of their cannon of evidence on amentia and heredity. In the sixth edition of his book on amentia, which was published in 1937, Tredgold introduced Vint's research on native brains in Kenya while discussing his findings on the cells in the prefrontal cortex of low-grade aments, in which he claimed that there were a greater proportion of immature cells. Tredgold went on:

> It is of interest to compare the above findings with those of an investigation into the condition of the brain of the East African native made by Dr. F. W. Vint. In this it was found that whilst the weights of 351 adult brains ranged from 35 to 57 ounces, the majority were between 43 and 47 ounces, and definitely less than the brain of the European. Histological examination showed that the thickness of the pyramidal cell layer (Lamina II.) of the prefrontal cortex was only 84 per cent. of that of the normal European ...[33]

Tredgold then went on to quote at length from Vint's 1932 article on the cell content of the East African pre-frontal cortex, where it described the preponderance of undifferentiated cells and the lack of pyramidal cells in the African cortex.[34]

Another eminent British specialist on amentia, R. J. A. Berry, who approached the subject from a mainline eugenic perspective, was also aware of the work of the Kenyan eugenicists. In Berry's 1933 edited book on mental deficiency, he cited the work of both Gordon and Vint, as well as his own research in Australia, and used the collective findings to conclude:

> This and other allied investigations into mental deficiency seem to suggest, if not indeed to prove, that mental deficiency is a manifestation of improper development and not of disease, so that the problem becomes one of preventive medicine, eugenics and embryology rather than of curative medicine. If, then, researches such as these should succeed in directing the attention of the profession and the lay public to the true causes of a great national problem, the labour and time devoted to them will not have been in vain.[35]

As described in the previous chapter, Berry also finished the same book with an article by Gordon which gave a eugenic interpretation of the Nakuru murder case.[36] Berry and Tredgold were among the most eminent experts on mental deficiency in Britain, and their use of Gordon and Vint's work as a part of the body of evidence on the innate, biological causes of amentia demonstrates how the Kenyan eugenicists had convinced important sections of the eugenic-minded medical establishment in Britain. Both men were, however, reaching the end of their careers in the 1930s, and represented a generation of experts on amentia that was to be superseded by the new thinking of the likes of Lionel Penrose. The simplistic hereditary approach favoured by Berry and Tredgold was thus beginning to be undermined just as they were taking up the theories of the Kenyans. The Kenyan eugenicists were ultimately working within the intellectual framework on mental deficiency that was becoming increasingly outdated; this was to become more apparent as the 1930s wore on.

The success of the Kenyan eugenicists in winning the support of significant figures in the British eugenics movement at the beginning of the 1930s is evidence of enthusiasm for research in the area of race and biology. This enthusiasm was partly caused by the new problematising of race in biological discourse. The increased uncertainty about the biological foundations of race meant that research which promised to examine race scientifically, and the most important social manifestation of possible race difference – intelligence – was seen as a valuable exercise. In the rest of this chapter, however, it will be shown that although there were significant supporters of the Kenyan eugenicists in the metropole, the problematic of race and intelligence was heavily disputed and there were powerful dissidents to the Eugenics Society's line in both scientific and official circles.

However, good relations between the Kenyan eugenicists and British Eugenics Society were maintained after the campaign for research funding faltered. Gilks was co-opted as a member of the council of the Society in 1934;[37] Gordon was elected to the Consultative Council of the Eugenics Society in 1939 and throughout World War Two he kept 'in close touch with the Society's headquarters'.[38] When Gordon died in 1947, an obituary was printed in the *Eugenics Review*. It merely described Gordon as 'keenly interested in all aspects of eugenics' and finished: 'Dr Gordon will always be remembered by us as a man of exceptional gentleness, courtesy and charm.'[39] Beyond mentioning his article 'Amentia in the East African', the obituary made no reference to Gordon's work on race and intelligence.

Race and intelligence in the British press

The debate in *The Times* about the Kenyan theories on race and intelligence that followed the Eugenics Society's letter is worth following closely because it helps unravel the issues about science and race that were under discussion among eugenicists and biologists in Britain in the 1930s. The first shot was fired when Gordon had a letter printed in *The Times* on 8 December 1933, in which he stated his and Vint's findings in a much more explicit manner than had been adopted in the Eugenics Society's letter of 25 November. Gordon included a graph comparing the brain capacity in cubic centimetres of Europeans with 'natives', which showed growth in African brain capacity gradually ceasing at adolescence. This was in contrast to the European brain, which was shown to rise steeply from about the age of sixteen. The peak of African brain capacity indicated in the graph was about that given for Europeans aged ten: 'It will be seen that the native curve of brain capacity ends practically where the European begins.'[40] Gordon also pointed out the correlation between his findings on brain capacity and Vint's findings on the cells in the prefrontal cortex. In this letter, the extremity of Gordon's position on race and intelligence was made clear.

The physical anthropologist Louis Leakey, who had spent much of his life in Kenya having been brought up there by his missionary parents, responded with a letter attacking Gordon's ideas. First of all he questioned the methodology used for measuring the cranial capacity of the subjects: he argued that the practice adopted by Gordon of making measurements on the living to estimate cranial capacity was notoriously unreliable and inaccurate. He then went on to argue that regardless of methodology, there was no basis for the premise that there was a correlation between cranial capacity and intelligence, pointing out that Neanderthal man had a much bigger brain capacity than modern man. Leakey questioned the source of the brains on which post-mortems were carried out, saying that if they were obtained from the prison, mental hospital or the infectious diseases hospital the subject could not be said to have been normal. Leakey attributed the cessation of growth in mental capacity at adolescence to the lack of stimulation in the normal conditions of African life and to the fact that sexual activity began at a younger age, somehow inhibiting mental development. These factors would be overcome by education and cultural development. Leakey concluded by arguing that Gordon had neglected to consider environmental differences – cultural, physical and dietary – which obstructed a real comparison between European and African intelligence.[41]

The psychologist Cyril Burt, who was well-known for his work on intelligence testing, also wrote to *The Times* to attack Gordon's approach, arguing that: 'Such conclusions are so contradictory to current theories and current practice that both the data and the logic on which they rest demand the closest scrutiny.'[42] Burt was Professor of Education at the University of London and psychologist in the Education Department of London County Council. He was also a longstanding member of the Eugenics Society. While teaching psychology and physiology at the University of Liverpool, he had performed intelligence tests at different schools: a preparatory school where most of the pupils were the sons of academics, Fellows of the Royal Society and clerics; an elementary school populated by the children of small traders; and a school in the slums. The boys at the preparatory school achieved the best results, followed by those of the elementary school, while those of the school in the slums did much worse. From this Burt concluded that intelligence was hereditary, with the obvious assumption that intelligence tended to follow social class.[43] Burt has recently become notorious after being exposed as producing fraudulent work on twins. He published three articles between 1943 and 1966 in which he claimed to have tested the intelligence of identical twins separated at birth, producing results that appeared to support a hereditary explanation of intelligence. Records could not be found of the twins tested by Burt and there was a suspiciously high correlation between the twins, suggesting that he fabricated his results.[44]

As an ardent believer in intelligence testing, Burt's criticism was based on a dislike of Gordon's use of measurements of cranial capacity as an indicator of intelligence. He also attacked the implication that Africans could not benefit from education:

> few scientists will accept the conclusions he has drawn – namely, that the natives of East Africa are incapable of rising to the standards of civilization, and that any effort to destroy their primitive simplicity by education along European lines may induce insanity.[45]

Burt particularly attacked Gordon's use of capacity as an indicator of intelligence. He considered Vint's research more suggestive, and also pointed out that Vint himself was not making the same bold generalisations. Burt commented that he shared Huxley's view that African races were probably slightly lower than Europeans in 'pure intelligence', but that the average difference between races must be small,[46] yet Burt attacked the Kenyan research, while Huxley defended it. The reasons for this were partly methodological: Burt saw intelligence testing rather than physiological examination as the most reliable measurement of intelligence. Also, as an educationalist, Burt recognised and distrusted the implications of the Kenyan research for educational policy in Africa. Burt's attack on Gordon is worth noting partly because he cannot be placed in the same camp as the likes of Hogben and Haldane. As a rather typical, old-fashioned eugenicist, who held the class assumptions that characterised conservative, 'mainline' eugenics, Burt's criticism is an example of how reactions to the Kenyan research cut across political camps within the eugenics constituency.

Huxley wrote a response in *The Times* that was ambivalent although ultimately defensive of the ideas put forward in Gordon's letter of 8 December 1933. Huxley conceded that until the overall average weight of the subjects of Gordon's measurements was taken into account, far-reaching conclusions could not be made, but argued that the slowing down in brain growth after puberty in Africans compared with Europeans was suggestive and that further work should be encouraged:

> Dr. Gordon's results make it clear that such an investigation should be undertaken at the earliest possible moment, for, *pace* Dr. Leakey, there does not appear to be a positive correlation between absolute (and probably still more relative) brain-size and intelligence, and *pace* Dr. Burt, a correlation of 0.3 though low, would make a considerable difference to the average of a population ... Should either native stature or still more relative brain-size lag behind those of Europeans from puberty onwards we should have a fact of great importance. If this proved to be correlated with poorer histological structure and with a falling off in performance as judged by really adequate intelligence and performance tests, we shall have to think of a revision of educational policy. But I agree with Dr. Leakey and Dr. Burt that the abler natives are perfectly capable of profiting by the best education we can give them.[47]

Haldane (who became Professor of Genetics at University College London in 1933), on the other hand, was quick to criticise Gordon's ideas about cranial capacity. He too wrote a letter to *The Times* on the subject, comparing Gordon's figure of 1,481 cubic centimetres as the average European capacity with that of Eskimos:

> Martin found that the average cranial capacity of a group of male Eskimos was 1,563 cubic centimetres. Charity begins at home, so let us hope that, if the natives of Kenya are to be protected from the European education, steps will be taken at the same time to safeguard Europe from the disintegrating effects of Eskimo culture, which may well prove too complex for beings of smaller cranial capacity.[48]

Hogben also strongly criticised Gordon's work.[49] Hogben was Professor of Social Biology at the London School of Economics and before that had spent four years in Cape Town in the zoology department; his experiences here 'did much to shape his abhorrence of racial prejudice and greatly influenced the direction of his career on his return to Britain'.[50]

Gordon's response to such criticisms was dismissive:

> I have never bothered to answer irresponsible critics like Haldane & Co (for
> this subject is not their province) who regard Vint & me (as each of us found
> to his horror in England – not in America & Germany) as 'bloody colonials'
> to be squashed at any price.[51]

He also argued that the measurements of brain size and weight, which
were the subject of much of the criticism, were not the most important
aspects of his and Vint's work.[52] It is true that the issues arising from
Vint's histological work in particular were not approached by these critics
in *The Times*. Haldane and Burt's criticisms were important in undermining the mental capacity campaign, but the fact that they failed to
address all aspects of Vint's research reveals something of the state of
discourse on scientific racism in the 1930s. The scientific debate
consisted of small-scale skirmishes around clearly biased excesses rather
than attacks on the premise of racial inequality itself. It was after World
War Two, when the extremities of Nazism came to light and the anti-
colonial movements became more forceful, that racism was more explicitly confronted in science, as both a moral and a scientific issue. In the
1930s, even important figures in making the first attacks on scientific
racism were not necessarily convinced that all races were equal: for
example even Haldane thought that although racial differences in intelligence had not yet been proven, he asserted that that did not necessarily
mean that 'the theory of absolute racial equality' was correct.[53]

There were less technical letters, though, which did question the
whole underlying notion of racial inequality in intelligence. One such
letter came from the Reverend John Levo, Chaplain to the Archbishop of
the West Indies, who wrote that in his experience, of 'pure negroes' in the
West Indies:

> Their mental development after puberty has been as full and rapid as that
> of white children similarly educated, and their record in manhood quite as
> satisfactory. This is, too, the considered judgment of the Archbishop of the
> West Indies, whose experience in the matter is unrivalled.[54]

Major G. Keane, former Director of Medical and Sanitary Services in
Uganda, also wrote testifying to the intelligence of Africans,[55] while Dr T.
Drummond Shiels, who had been parliamentary Under-Secretary of State
for the Colonies under the Labour government of 1929–31, contributed a
letter that directly confronted and indicted the political implications of
the Kenyan work:

> The recent correspondence in your columns on the subject of 'The Native
> Brain' has more Empire significance than has perhaps been realized. It is
> to be hoped that the weighty letters of your scientific contributors have
> given pause to a number of Dr Gordon's listeners and readers who too
> readily assumed – from the account of an admittedly small and inadequate

investigation – the mental inferiority of millions of our African fellow subjects ... Africa – for us and for other Colonial Powers – has problems enough without the complication of rash generalizations on the relative mental capacities of white and black. What the ultimate truth of this may be does not affect the line we ought to take politically and socially. It is to give opportunity for the African to rise to his fullest possible stature, whatever that may be, by the provision of adequate educational facilities and by the granting of social and political responsibility according to capacity, tested by progressive advance. The proof of the brain is not in its anatomy or histology but in its product.[56]

These letters, although not engaging with the science of the Kenyan research, point to a growing wider culture of discomfort and doubt about the idea of racial inequality which was also to be crucial in shaping official nervousness about the Kenyan eugenicists.

John Gilks wrote to *The Times* in response to these critical letters. He pointed out that Gordon's findings were preliminary and had been published in order to put the case for a more comprehensive inquiry into backwardness, and claimed that Gordon and Vint's studies had been executed independently. The fact that their results corresponded so closely reinforced the case that the research was on to something and justified fuller investigation. Gilks finished by appealing to Britain's imperial duty:

> The mental capacity of backward races for whose welfare the British Empire has assumed responsibility has long been the subject of acute controversy; surely it is time that the matter be settled by the accumulation of facts. If our administrative or educational systems in East Africa require modification, it is infinitely better that this take place now while the country is in a comparatively early state of development rather than as a result of a long and costly system of trial and error.[57]

Gordon also defended himself in a letter printed in January 1934 in which he argued that the criticisms that had been made of his measurements of brain size and weight were missing the most important point of his argument, which was about the lack of African cerebral development after puberty.[58]

The correspondence on 'native brains' then stopped until August 1934, when Sir Ernest Graham-Little, an Independent Member of Parliament, whose constituency was the University of London and who was the most active parliamentary supporter of the Kenyan research, renewed the debate. Graham-Little argued that research into the causes of 'native backwardness in mental and physical development' was urgently needed in the interest of native welfare.[59] This letter was quickly followed by letters in agreement with this point of view from Elliot Smith and Lord

Dawson.[60] Graham-Little provoked a cautionary response from Cullen Young of the Religious Tract Society, who had been a missionary in Nyasaland until 1933. Young disputed the notion that the onset of senility could be expected from the age of thirty-five in the African, and warned of the danger of making generalisations that 'were demonstrably false regarding "the African native" as a whole'.[61]

The arguments in the correspondence columns of *The Times* also prompted an editorial on the subject, in which it was contended that an enlarged research project should be funded in order to establish the truth that lay beneath such controversy. The editor acknowledged the criticisms of Gordon's research but argued that the question was still open and that the intentions behind the Kenyan research were admirable:

> Dr Gordon suggested that European methods of teaching may not only be harmful; but he did not suggest, as some of his critics appear to have supposed, that no education ought to be given. His plea was for means adapted specifically to the ends and objects in view. Such adaptation, as he urged, is impossible without a much more extensive knowledge of native psychology and of the structure and capacity of the native brain than is now anywhere available. It is mere cruelty to inflict upon one kind of mind the teaching suitable to another and different kind of mind; nor does there seem to be ground for the assumption that, because Europeans and natives are partners in the same human family, therefore what is good for the one must necessarily be good for the other.[62]

The Times was traditionally in favour of eugenics; the passing of the 1913 Mental Deficiency Act was partly attributed to the publicity given to the subject by the newspaper. The response of other publications to the Kenyan ideas on race and intelligence will now be considered.

In 1932 many British newspapers covered the story of the Kenyan research after Sequeira succeeded in publicising the matter in an article in the *British Medical Journal* (*BMJ*). The *East African Standard* recorded the reaction of some of these mostly local newspapers, giving an interesting account of lay responses to scientific racism.[63] The *Observer*, the *Lancashire Daily Post*, the *Glasgow Herald*, the *Yorkshire Post*, and the *Norwich Eastern Daily Post* reported on the research, with the latter two devoting leading articles to the subject. All these newspapers agreed on the significance and utility of the work, although most warned of the dangers of drawing hasty conclusions from preliminary findings. Despite such words of caution, the social and political ramifications were seized upon. As the editor of the *Yorkshire Post* put it, the research suggested that changes would have to be made to the methods of administration and education Europeans established for East Africans.[64] The scientific authority of Sequeira and Vint was clearly important in justifying the seriousness with which the Kenyan research was treated in the British non-medical press.

British medical journals also gave the Kenyan research quite extensive coverage. In November 1933 the *Lancet* merely covered the story without providing much commentary on the subject beyond expressing the hope that Gordon's plea for funding would find a sympathetic hearing,[65] but in February 1934 a more critical editorial was published. This leading article not only attacked the most obviously flawed aspects of the Kenyan research that had been highlighted by Haldane, Leakey and Burt, such as the limited value of measurements of brain size and cranial capacity. Vint's microscopic work, which experts had hitherto been reluctant to criticise, was also given short shrift. The editorial concluded by attacking the notion of the early degeneration of the African brain:

> In the absence of enough acceptable intelligence tests it cannot be asserted that there is a loss of intellectual ability; the native boy who has hitherto been a 'bright, malleable, nice little fellow' and who after 15 disappoints his employers by failing to develop as they had hoped, may have regressed in usefulness in the circumstances in which nature and colonial expansion have combined to place him, but cannot, without further proof be regarded as having regressed in intelligence.[66]

Most of the criticism that had been produced before this had focused on Gordon's research rather than Vint's. This is only partly explained by the fact that Gordon had created more publicity for himself by visiting Britain to publicise the research and by writing to *The Times* because the work of Gordon and Vint on race and intelligence were usually treated as one. The reason for this emphasis on Gordon's research lay in the transparency of Gordon's failings. Vint's histological research was more specialised and the problems of such methodology had not been so vocally rehearsed, particularly in relation to race. The editorial in the *Lancet*, therefore, represented quite a bold stance against the Kenyan eugenicists.

The *BMJ* gave the Kenyan eugenicists more column space. It was the journal of the BMA, which contributed £50 to Dr Gordon for research on amentia amongst the natives of East Africa as a part of the Science Grants awarded for the period 1932–33.[67] Despite this gesture of support, the *BMJ* was clearly ambivalent about the Kenyan research. In March 1932 Sequeira wrote an article in the *BMJ* on the subject in which he summarised Gordon's work on cranial capacity and Vint's on cell structure and argued that 'the physical basis of "mind" in the East African' differed from the European. Sequeira warned that the introduction of education along European lines might have disastrous consequences, just as it was inappropriate to educate backward and defective children using normal methods.[68] A leading article (the editor of the *BMJ* at this time was Norman Horner) appeared the following month which cautioned that the conclusions drawn in Kenya were premature and overbold, and that some of the biological precepts on which the arguments were based could not be accepted as established fact. Another major concern expressed in the editorial was the lack of tribal distinctions in the research and the definition of the East African as being in contrast with 'negroids'.[69] In the same issue of the journal, Dr Norman Leys, the famous humanitarian defender of African interests in Kenya, had a letter published in which he dismissed Gordon's conclusions on cranial capacity on the grounds that he failed to relate the figures to the overall size and body weight of the subject.[70] But the following year, after Gordon's trip to London in which he presented his work on 'Amentia in the East African Native' to the Eugenics Society, a less critical editorial was printed in the *BMJ*:

> A difference in biological level may well be the foundation of racial backwardness and be contributed to by both nature and nurture. None will disagree with the conclusion that wide team research is a 'pressing necessity'.[71]

The *BMJ* consistently reported the progress of the campaign by the Kenyan eugenicists and the responses given by the British government until 1936. The general approach of the journal was that the research should be continued and expanded in order to determine the uncertain role of environmental and hereditary causes.

The Kenyan research clearly aroused quite considerable interest and controversy in Britain; the debate on race and intelligence drew on different issues, ranging from methodological disputes to general arguments about the equality of races. In the 1930s the underlying principles, practical and moral, of scientific racism were under discussion as a result of the growing sophistication of understandings of genetics and the role of environment. The controversy over the Kenyan research is a chapter in the history of British scientific racism which exposes the breadth of issues and the wavering constituencies involved in its demise.

Official interpretations

Staff in the Colonial Office in London, meanwhile, followed this debate about African intelligence with growing distaste. The Colonial Office was initially quite tolerant of the Kenyan eugenicists, receiving their applications for funding and support with either genuine engagement or, more often, polite but fundamentally unimpressed interest depending on the individual concerned. As the full implications of the Kenyan eugenics programme unfurled, attitudes hardened and the Kenyan proposals were rejected.

The personnel of the Colonial Office in London were less amenable to the ideas propounded by the Kenyan eugenicists than that of the previous Conservative government may have been. Leopold Amery, Secretary of State for the Colonies between 1924 and 1929, recommended a eugenics programme in 1931 to counteract the 'short-sighted sentimentalism' of British social policy.[72] He was generally pro-settler in his attitude to Kenya and averse to native paramountcy, one of the great subjects of debate on Kenya since the Devonshire Declaration of 1923, which stated that native interests in Kenya should be paramount over those of European and Asian settlers. Amery was also a close friend of Edward Grigg (Governor of Kenya, 1925–30), who was a passionate supporter of Gordon's research.[73] In the late 1920s Kenyan eugenicists would probably have found a more responsive and enthusiastic political environment than they did in the early 1930s. Tension between settlers and the British government rose in the 1930s. The Labour government of 1929–31, with Lord Passfield (Sidney Webb) as Secretary of State for the Colonies, had made settlers furious with its *Report of the Commission on Closer Union* and *Memorandum on Native Policy*, in which settler self-government was repudiated and native interests reasserted.[74]

Just before the Labour government fell in 1931 in the face of economic depression, a further report, that of the *Joint Select Committee on Closer Union*, was published.[75] In this report, the idea of Closer Union (of the British East African territories) was finally laid aside, along with settler hopes for self-government, while the issue of native paramountcy was put to rest through a reinterpretation that was more acceptable to the settlers. Paramountcy became reconciled with and essentially reduced to the dual policy based on Rhodes' conception of 'equal rights for all civilised men', which enabled the settlers to uphold their supremacy. Although the Labour government crucially failed to establish native paramountcy, the renunciation of settler self-government changed the agenda in the dialogue between settlers and the Colonial Office in the 1930s.[76]

In 1931 the conservative-dominated National Government came to power, headed by Ramsay MacDonald between 1931 and 1935. The new Governor, Joseph Byrne (1931–37), and Lord Francis Scott, Kenya's most prominent settler following the death of Lord Delamere in 1931, attempted to improve relations between the British government and settlers. However, settler relations with the Colonial Office deteriorated, first of all over the introduction of income tax, and then over land, when gold was discovered in a native reserve at Kakamega.[77] A settler delegation on the matter of income tax, led by Grogan, left the new Secretary of State for the Colonies, Cunliffe-Lister, with a 'most unfortunate impression'.[78] The discovery of gold at Kakamega in the midst of economic depression caused great controversy in Britain. Settlers flocked to Kakamega and in order to gain access to the (vainly) hoped-for gold, the affected land was removed from the reserve through a 1933 amendment to the Native Lands Trust Ordinance of 1930. Cunliffe-Lister allowed this to go ahead despite the betrayal of trust it represented to Africans over the crucial question of land.

Kakamega, with the question of African rights and land security that it brought up, mobilised the humanitarian lobby and caused enormous controversy in Britain. Some 174 articles appeared in the press following the outbreak of the controversy in December 1932. In February and March 1933, thirty-two questions were asked on the matter in the House of Commons, mainly by Labour MPs, and the Colonial Office received nearly 300 letters and petitions. Those involved in the campaign included Ormsby-Gore (Secretary of State for the Colonies, 1936–38), the Archbishop of Canterbury, the Church Missionary Society, the Church of Scotland Mission, the Society of Friends, the Anti-Slavery and Aborigines Protection Society and the Negro Welfare Association.[79] Therefore the campaign by the Kenyan eugenicists in Britain occurred at a time when the settler treatment of Africans in Kenya was a live issue in the metropole. The Colonial Office had permitted the amendment of the Native Lands Trust Ordinance, but the disapproval this had elicited sensitised them to the political implications of settler claims in the form of eugenic arguments.

Despite this, in the early stages of the campaign, officials were fairly responsive to the proposals of the Kenyan eugenicists; central to this early success was the effectiveness with which its proponents related the research to current metropolitan debates on African development. Since the late 1920s, metropolitan interest had developed in systematic research into the conditions, welfare and psychology of Africans since the advent of colonial rule. It was considered necessary by members of the Colonial Office and others concerned with 'native welfare' to institute a policy of welfare and modernisation in Africa based on up-to-date knowledge of African conditions. Hetherington has described this concern with development and research as an important aspect of the 'paternalist faith' of the inter-war period, which placed a new emphasis on 'Britain's role as that of a kind of development agency'.[80] By the end of the 1930s, development came to mean 'a process of social, political and economic change' directed by Britain 'so that African territories could be transformed into viable states'.[81] Pearce on the other hand has argued that the period before 1938 was one of 'complacent trusteeship' and that it was only in the decade 1938–48 that a stagnant colonial policy was shaken off and serious consideration was given to African development with a view to eventual self-government.[82] Pearce mentions Gordon's research as an example of the old-fashioned racist views that were 'commonplace in the inter-war period but did not go unchallenged'.[83] Pearce is correct in that the views of the Kenyan eugenicists were eventually rejected and superseded by new perspectives on backwardness in Africa and how to combat it. However, there was greater continuity in the discourse on development between the Kenyan eugenicists and those concerned with colonial development than is appreciated by Pearce; this continuity was expressed in the metropolitan movement for research into African development and education that began in the mid-1920s.

An early proponent of the need to establish a research survey focusing on African welfare and development was Oldham, Secretary of both the Conference of British Missionary Societies and the International Missionary Council and an active member of the Secretary of State's Advisory Committee on Native Education in Tropical Africa (later to become the Advisory Committee on Education in the Colonies). The advisory committee contained missionaries, education officers, representatives of the Colonial Office and educational experts, and had an important role in formulating colonial educational policy. Oldham was at the centre of a group of British humanitarians that included Norman Leys, Leonard Woolf, Macgregor Ross, Randall Davidson (the Archbishop of Canterbury) and Sir Humphrey Legett, all of whom were particularly concerned with native interests in Kenya. The Kenyan settlers had acquired notoriety among British liberals and humanitarians after such incidents as the flogging in 1907 of three allegedly disrespectful African rickshaw drivers outside the Nairobi courthouse by the prominent settler Grogan and two others, cheered on by a crowd of settlers. 'Pro-native' awareness of the matters of land and labour had been aroused early in the colony's history by the Maasai moves, which were completed by 1913; the 1919 labour circular produced by Ainsworth was also important in attracting the attention of British humanitarians to Kenya. Oldham and Randall Davidson organised a memorandum entitled 'Labour in Africa and British Trusteeship' which attacked labour practices in Kenya and was an early call for 'native paramountcy', signed by sixty-nine prominent British humanitarians. Organisations like the Anti-slavery and Aborigines Protection Society and the Fabian Society also continued to keep pressure on issues relating to native welfare in the inter-war period.

Sidney Webb described Oldham as the 'chief wirepuller' of the time;[84] in the late 1920s, Oldham increasingly turned his attention to the pursuit of a large-scale research project into African conditions.[85] In 1925 he put forward the idea of examining African conditions and welfare needs; this was discussed at the East African Governors' conference of 1926. From this Edward Grigg concluded that £10,000 per annum should be spent on researching the East African native and the effects of European civilisation.[86] Grigg's idea did not come to anything and the Colonial Office eventually dismissed it as 'vague and woolly'.[87] Despite this, Oldham reported in 1927 that he had found Amery, Lord Lugard and Balfour enthusiastic about the idea of research into native affairs.[88] It was the desire to record and analyse the state of the African peoples under the influence of colonial rule and its changes that led to the foundation of Hailey's *African Survey*.[89]

At the time that the need for African research was first being posited, around 1926, Gordon started giving lectures in Kenya on the subject of 'The Mind of the Native'. The *East African Standard* reported extensively on one such lecture, and the suggestions and ideas put forward by Gordon appeared very similar to those expressed by Grigg and Oldham.[90] Dr Gordon's more grandiose claims about the political and the social implications of his research for colonial trusteeship were conflated with the ideals of a more general African research programme by Grigg, who enthusiastically supported Gordon's ideas. In response to the research proposal suggested at the Governors' conference, an editorial in the *East African Standard* reported on the relevance and significance of Gordon's theories:

> If it is considered necessary to investigate the reasons why sleeping sickness or rinderpest, cotton boll worm or mealie bug exist, surely it is much more essential to study the mind and the mental processes of ten millions of negroes in East Africa upon whose proper development the whole future of the Empire will depend.[91]

Before the researches and ideas of Gordon and Vint were formulated and elaborated in the early 1930s, it had therefore already been suggested that money should be invested by the British government in researching the African population and Kenya had been isolated as an appropriate area for such a study by Oldham and, in particular, Grigg.[92]

The survey of African conditions proposed by Oldham was to examine many aspects of native life: access to land, agricultural production, demographics and public health.[93] In addition to these social and economic interests, African educational responses and needs were considered crucial. There was a strong feeling among those interested in education in the African colonies that it had to be appropriate to African conditions, and that European educational policy and method should not be simply transplanted in an unmodified form. This approach can be seen in the 1925 memorandum on education produced by the Advisory Committee on Native Education: 'Education should be adapted to the mentality, aptitudes, occupations and traditions of various people, conserving as far as possible all sound and healthy elements in the fabric of their social life'.[94]

The idea that education needed to be tailored to Africans brought up the issue of racial differences in educable capacity. Dr Keppel of the Carnegie Corporation, who already had an interest in promoting education for African-Americans in southern US states, was struck during a visit to East Africa by the need to provide educational material that was more suitable for Africans.[95]

In order to create appropriate educational material, Keppel and Oldham agreed it was necessary to study African mentality; in 1932 the Carnegie Corporation sent Richard Oliver to perform intelligence tests at African schools in Kenya, which are discussed in Chapter 6. In around 1930, Arthur Mayhew, who after twenty years' experience in education in India was Joint Secretary of the Advisory Committee on Education,[96] wrote a memorandum on the possibility of introducing the testing of mental capacity to the colonies. Mayhew quoted from the 1924 Report of the Consultative Committee of the Board of Education, which emphasised the value and use of intelligence testing for educational purposes, subject to limitations imposed by the fact that the science of intelligence testing was in its infancy. Mayhew went on to recommend the introduction of testing to colonial education for the purposes of detecting mentally defective children or particular vocational aptitudes, the classifying of pupils on entry into school and the discovery of 'important points of difference between the European and non-European races in the various modes of intellectual reaction'.[97] Therefore, the examination of mental capacity in Africans was not of exclusive interest to the Kenyan eugenicists; intelligence testing had a respected role in British educational policy at this time, which was taken up by individuals in the metropole who had a role in shaping colonial educational policy.

After Dr Gordon had his letter published in *The Times* of 8 December 1933 (described in detail above), which received mention on the front page and revealed the full extent of Gordon's appraisal of racial differences in intelligence,[98] Mayhew wrote to Oldham about the 'Gordon bomb'. Gordon's letter was shocking in its racial comparison of brain capacity, implying substantial African inferiority in relation to the European. Mayhew wrote to Oldham saying 'I cannot help seeing in it signs of a reactionary movement that is spreading over East Africa with a view to limiting educational aims and enterprise'.[99] Unfortunately, Oldham's response has not survived, although from the next letter from Mayhew, it would appear that Oldham shared Mayhew's disapproval.[100] Oldham's correspondence with Matheson, the Secretary of Hailey's African Research Survey, also indicates that Oldham was uneasy about Gordon's letter.[101] A distinction clearly needs to be made between the position of the likes of Oldham and Mayhew and the ideas of the Kenyan eugenicists. Oldham and Mayhew approved of the introduction of intelligence testing because it might help in the detection of racial differences in mentality but were horrified by the extremity of Gordon's statements on race and intelligence.

The first attempt to obtain British funding for the Kenyan research was made in 1931 when Gilks submitted to the Kenyan government an application for a grant from the Colonial Development Fund (CDF). The proposal, however, was not sent on to Britain as there were at that time so many applications being made and no further action was taken.[102] This original application to the CDF required a capital expenditure of £2,255 and an annual recurrent expenditure of £8,700. The research was expected to take three years and its objective would have been to form the foundations on which educational policy could be based. The research plan was to carry out both intelligence testing and physical measurements of different African tribes, and to pursue Vint's histological examination of African brains.[103]

Dr Sequeira started the campaign in Britain in earnest in the spring of 1932, when he visited Sir David Munro of the Industrial Health Research Board of the British Medical Research Council, and his response was positive. Munro wrote to Dr Stanton, who was Medical Adviser to the Colonial Office, about his meeting with Sequeira in which he stated that 'the work of Gordon and Vint is obviously on a subject of considerable importance'.[104] Munro also wrote to Gordon offering the help of Eric Farmer, who had suggested earlier research into African educable capacity, and who was also with the Industrial Health Research Board.[105] The interest of the Industrial Health Research Board in intelligence lay in the desire to ascertain mental and physical ability with a view to occupational recruitment and training. It was felt by Munro that:

> This suggestion falls in with the general trend of colonial policy to extend imperially the study of industrial health problems as part of the general problem of medicine and hygiene in the tropics. It is thought that the Colonial Office would probably agree that in past days neglect of the study of such problems has contributed towards some unfortunate features in the development of native civilisation in other parts of the African continent.[106]

During his trip to London, Sequeira also started to recruit the support of the likes of Arthur Keith and Grafton Elliot Smith, who were encouraging about the importance of such a research programme. Sequeira also visited Professor Berry at Stoke Park in Bristol and Dr Tredgold, both experts on mental deficiency, and saw Stanton and Dr Anderson, acting secretary of the BMA.[107]

Dr Stanton recommended to Sequeira that Gilks write a proposal outlining the research programme. Gilks' resulting scheme for the study of native mental capacity and 'brain equipment' included a team consisting of a histologist (Vint), one or preferably two experimental psychologists, a psychiatrist, an anthropologist, laboratory assistants and a social worker. Gilks argued that Vint's preliminary findings proved beyond all doubt that a histologist was necessary; in other words Vint's research succeeded in legitimately placing (in the eyes of Gilks and his fellow campaigners) the idea of biology at the centre of this programme of researching African psychology.[108] The programme, which was devised by Gordon, Vint, Scott, Kauntze and Gilks, was planned to include the continuation of Vint's histological work, combining it with a psychologist who would devise a mental test to ascertain 'native norms', assess individuals' vocational capacity and estimate the aptitude of the population. The psychiatrist would diagnose mental disorders and classify mental deficiency, and establish the incidence of mental disease. The psychiatrist was also expected to examine the relationship between mental disease and educability, and in particular to see if mental disorders were aggravated by the attempt to educate Africans. The anthropologist was required to work out the relationship between native culture and education. A social worker was described as 'essential to help the psychiatrist and psychologist and to study in the homes of the people the questions raised by investigators'.[109] Richard Oliver was to be the experimental psychologist. The advantage of Oliver was that he had already worked on intelligence testing in Kenya for two years, was keen to continue his work and was happy to join the team. What is more, Oliver was funded by the Carnegie Corporation, and his appointment had been approved by Oldham.[110] The programme was estimated to require £4,062 a year for new salaries and other expenditure such as travel, apparatus and clerical staff. Gilks wrote that the work would have to go on for at least two years to get significant results.[111]

The early response to Gilks' letters was fairly promising. Munro was keen on the ideas; he wrote: 'I think the time is ripe, or at any rate ripening, for some sort of Conference on these African proposals. It would be a good thing, I think, if the Colonial Office took the initiative.'[112] The Colonial Office, however, as well as being concerned by its expense, recognised the wider implications of such a research programme: 'I pointed out to Munro that the proposals, as a group, if pursued might have reactions in the political sphere.'[113] It was suggested that the International Institute of African Languages and Cultures might be a more appropriate source of funding and support, but when Oldham, who represented the Institute in such matters, was contacted he said that he was unable to give special help, although the men sent to Kenya by the Institute for other reasons could assist.[114] Dr Yates, an anthropologist funded by the Institute of African Languages and Cultures, arrived in Kenya a few months later, and it was expected that his work would dovetail with the work of the other disciplines in Kenya working on intelligence.[115] There is, however, no further mention of Yates in relation to the work on mental capacity.

Gilks, however, was greatly encouraged by the response from the Colonial Office and wrote to Stanton again at the end of September asking about the possibility of funding to keep Vint on to do laboratory work.[116] Stanton's reply was that the Colonial Advisory Medical Committee would be inclined towards supporting a proposal for research with a grant from the Tropical Diseases Fund, and that provided the proposal was considered to be sound, the Fund would give Vint a grant of up to £250 to maintain his laboratory work.[117] Meanwhile, Gilks was spreading the word, reporting to Stanton that his research proposals were being enthusiastically received at meetings he had held both in South Africa and in Nairobi.[118]

Despite Gilks' optimism, even at this stage in the campaign, reactions in the Colonial Office were sceptical and sensitive to potential criticisms. Stanton wrote in the comments written on the file:

> The question is only to a limited extent a medical one but I am advised by my specialist colleagues that on the basis of scanty data, rather sweeping conclusions have been reached in Kenya … From my point of view the proposals are rather fanciful and I am not inclined to support them on the medical side. There are political possibilities associated with the proposals which will not have escaped your notice.[119]

Others made similar comments, arguing that there was sufficient doubt about the real scientific value of the work and sufficient potential for political complications to make it undesirable for both the Colonial Office and the Kenya government to be too closely associated with the proposals or to provide funding. For these reasons, it was seen as better for the suggestions to be passed on to a non-governmental institution such as the International Institute of African Languages and Cultures.

Sequeira's and Gilks' activities in 1932 seem to have petered out without much progress in the form of financial aid, although the support of important British scientists had been obtained. The situation was then left until October 1933, when Dr Gordon visited Britain. During this trip, Gordon presented his paper to the Eugenics Society; he also addressed the Empire Parliamentary Association, a non-party political, unofficial parliamentary body interested in imperial affairs. Its vice-presidents in 1934 included Ramsay MacDonald, Cunliffe-Lister, Amery, Lord Passfield (Sidney Webb), George Lansbury (Leader of the Opposition), Stanley Baldwin and Winston Churchill.[120] Despite its unofficial status, the Empire Parliamentary Association clearly presented a politically influential audience for Gordon's ideas. Unfortunately, the records of the

Association do not record any details of Gordon's speech or its reception. Following the Eugenics Society's letters to *The Times* of 25 November, a question was raised in the House of Commons on the subject of the Kenyan research. Captain Erskine-Bolst asked Sir Phillip Cunliffe-Lister, Secretary of State for the Colonies, whether he proposed to take any action on the matter. Cunliffe-Lister replied that it seemed unlikely that the Kenyan government could afford to fund such research, and that he was reluctant to suggest that the British government should do so.[121]

Gordon did, however, visit the Colonial Office to present the case for funding. In support, Dr Paterson submitted a proposal for research into the East African capacity for mental development, which involved the expenditure of £2,255 in capital and £8,700 annually for three years.[122] Gordon also discussed the possibility of acquiring funding from the EAC. Hemming, of the EAC, was sympathetic to Gordon's proposals, telling Gordon that he had had informal discussions with the Colonial Office on the matter.[123] Moore (Colonial Secretary, who was Acting Governor between 11 June and 17 November 1933 while Byrne was on sick leave) then wrote to Flood, Assistant Secretary in the Colonial Office, saying the Kenya government would 'gladly support' the proposals, although it was unable to make a financial contribution itself.[124] As a result of Moore's support, there was now more of a basis for activity from the EAC. Hemming's idea was that provided the Colonial Office was willing, the EAC could arrange an 'informal conference' on the matter. Its purpose would be to consider whether the Prime Minister, with the approval of the Secretary of State, should be urged to set up a small EAC committee which would report on the scientific importance of the work and suggest means of obtaining funds for it.[125]

Flood replied to Moore stating that he had told Gordon that he was 'in full sympathy with his ideas', but that there was no prospect of getting funds from Kenya or the British taxpayer.[126] Despite such assurances of support for Gordon's ideas, Flood did in fact have serious doubts about Gordon. Along with Thomas Stanton, Chief Medical Adviser to the Colonial Office, Flood was one of the main Colonial Office officials who dealt with the Kenyan eugenicists. Berman has described Flood as a 'vigorous critic of Kenya policy and the settlers'.[127] As mentioned above, as early as 1932, Stanton had expressed some suspicion about the Kenyan research presented by Gilks, pointing out that 'sweeping conclusions' had been made on 'scanty data'.[128] The Colonial Office was also quick to appreciate the political implications of such theories.[129] The somewhat qualified support for the Kenyan eugenicists within the Colonial Office evaporated more swiftly as a debate began to rage in the British press in the winter of 1933/34 about 'the native brain'.

Gordon started to question the sincerity of the Colonial Office's interest in the proposals after Flood had informed him that money would not be given by the British government.[130] To publicise the issue, Gordon wrote his provocative letter to *The Times* of 8 December 1933.[131] This letter and the debate it stimulated caused the Kenyan eugenicists to be more clearly identified with settler political interests and racial hostility. Flood wrote:

> He [Gordon] has now burst into *The Times* with a letter on the subject of the mental capacity of 'the native' and has drawn comparisons which are very much to the disadvantage of the said native as compared with the European. This probably means that the question will be considerably ventilated by the various people who take the view that the black man is a lower order of humanity altogether ...[132]

The critical responses to Gordon published in *The Times* were noted by the Colonial Office. Haldane's remarks certainly influenced the Colonial Office in their discussions on the value of the theories on the East African. In a 1936 memorandum which went over the history of the campaign for funding by the Kenya doctors, Flood said of the question of intelligence and brain capacity:

> an end was put to that by Professor Haldane, who pointed out that the Eskimo's brain was bigger than that of the European without apparently any corresponding results in intellectual power.[133]

As the Kenyan eugenicists became more active and determined on the subject of their research programme, Colonial Office suspicion about their methods and their aims grew:

> Dr Gilks called: said that there was a strong movement in Parliamentary Circles 'in support of the enquiry' ... I regret that, from his peculiarly furtive expression and demeanour, I have formed the impression that if there is any 'ballyhoo' in Parliament he is at the bottom of it ... it appears that he has always been an enthusiast on this particular line.
> I have now ceased to trust him.[134]

Officials in Nairobi, however, continued to support the research. As Head of Medical and Sanitary Services, Paterson had written a memorandum on the subject to Wade (Sir Armigel Wade became Chief Secretary in 1934, officiating as Acting Governor from March to September 1935 while Byrne was on sick leave) promoting the cause and arguing that the progress of African races depended on ascertaining levels of mental capacity and the causes of mental backwardness.[135] Byrne wrote to Cunliffe-Lister about Paterson's memorandum: 'With the conclusions reached in the memorandum I am in general agreement and have no doubt that the institution of research on these lines would help to solve many of the administrative, educational, social and political problems'.[136] The expressions of support for the research from the Kenya government, in each case in response to memoranda written by Paterson, clearly had some impact on the Colonial Office in London in taking the matter more seriously.

The statement of support from the Kenya government seems to have had an effect. It was decided that, depending on the approval of the Prime Minister, a special committee of the EAC would be formed, to which the government of Kenya would present evidence for the case that East Africa would benefit from the research programme.[137] The Colonial Office wrote to the governments of Nyasaland, Northern Rhodesia and Somaliland to obtain evidence from them to present before this proposed special committee.[138] Paterson also addressed the Colonial Advisory Medical Committee in October 1934 and reiterated the significance of research into African backwardness.[139] Despite the doubts of Colonial Office officials, an important step was about to be made: discussions were already being held about the composition of the proposed committee of the EAC, with women's groups writing to argue that female experts in the relevant professions be represented.[140] Hemming was favourably disposed to the research plans and was quite optimistic about the possibility of the provision of funding. Flood, however, was deeply sceptical about the value of such a committee and whether it would reach any meaningful conclusions, let alone produce any funding.[141]

The terms of reference for the proposed committee were:

> To examine the existing state of knowledge regarding the influence of brain growth and other factors upon mental development among the indigenous races of East Africa; to consider whether further research into this subject is desirable, and if so, in what direction it should be pursued, and to suggest means to that end.[142]

Despite the fairly restrained tone of the terms of reference, the Prime Minister, Ramsay MacDonald, had misgivings about the government being connected with the proposed research. Helen Tilley has shown how MacDonald consulted Oldham on the Kenyan research, who at this point came out strongly against it.[143] It was suggested instead that the question be referred to Sir Malcolm Hailey as a part of his survey into Africa; the Prime Minister agreed to this, but apparently rather reluctantly – his feeling being that such work would best be undertaken by a completely independent body such as the Rockefeller Institution.[144] The Prime Minister was particularly anxious that government involvement might have political ramifications in South Africa.[145] The Secretary of State for the Colonies agreed and it was decided to postpone the proposed committee and any enquiry about brain growth and mental development until Hailey had the opportunity to correlate its importance with other issues related to conditions in Africa.

Despite these rebuffs from the highest levels of the British government, the Kenyan eugenicists persisted in their attempts to acquire funding for an inquiry into racial backwardness, this time with the support of Graham-Little. At the combined meeting of the East African branches of the BMA in Kampala in May 1936 a set of resolutions was unanimously passed urging an immediate inquiry into the causes of mental and physical backwardness in East Africans. A copy of these resolutions had been sent to Ormsby-Gore, Secretary of State for the Colonies, and Graham-Little asked him in the House of Commons whether action would be taken. Ormsby-Gore's response was that no further action could be proposed until Hailey's survey was published.[146] Graham-Little was not satisfied by this and in September 1936 he wrote a controversial letter, published in the *BMJ*, which accused the Colonial Office of inaction:

> The pretext that the Colonial Office is waiting for the publication of a wholly unofficial report [Hailey's survey] should no longer be allowed to delay the action which all qualified observers are agreed is urgently and immediately called for.[147]

In a memorandum written in response to this, Flood maintained that finance for such a project would have to come from the government of Kenya, and that if Kenya did have £8,000 to spare for the project, it should be spent to much greater immediate advantage on essential medical and public health services.[148] As Flood, by now utterly sceptical of the Kenyan eugenicists, commented: 'I am not at all sure that Nairobi, which is in Kenya; in the Highlands; among the worst kind of native; and with the most poisonous kinds of whites; is at all the proper place for such a centre.'[149]

The controversy over the research into mental capacity became more intense at this time because it then became confused with the question of Vint's career prospects. It seems that for several years Vint had been applying for promotion or transfer and, according to the Colonial Office, had become tired of Kenya and had decided that if he were not given a promotion he would retire and take the gratuity. Vint was then offered the post of Senior Pathologist in Mauritius and was prepared to accept it. Many Kenyan doctors, largely those who had been involved in the mental capacity campaign, were furious and, with the support of the Governor, argued that to lose Vint would be a disaster for research in Kenya. Matters were made worse by the fact that the medical research laboratory in Nairobi where Vint worked was felt to be under pressure since the director of the department, Kauntze, had been promoted to Director of

Medical Services in Uganda and had not been replaced, and Dr de Smidt, the next in command, had been laid off after losing his sanity.[150] At the suggestion of the Governor, the Medical Research Council was approached for a grant to supplement Vint's income, but it declined to use its funds for such a purpose.[151] Vint's failure to obtain a promotion within Kenya was interpreted in the colony as part of a conspiracy against the research into mental backwardness. On 11 November 1936, an article published in the *East African Standard*, entitled '"Medical Crisis" in East Africa', suggested that the real reason for Vint's transfer was the hostility of the Kenya government to his work and a further letter from Graham-Little to the *BMJ* was reprinted.[152] In this letter Graham-Little argued that the depletion of the Medical Research Laboratory in Nairobi, particularly if Vint were to go, had resulted in 'a really urgent crisis now developing in the medical service of East Africa'.[153] This letter succeeded in annoying the Kenyan doctors Graham-Little was attempting to support, as in fact successive governors had supported the laboratory and it was still running quite efficiently.[154] In the end, Vint remained in Kenya. This was following a petition to the Governor, Joseph Byrne, composed by Kenyan doctors protesting against his transfer. This petition was signed by many of the longstanding supporters of the research into backwardness.[155] Byrne sent a telegram[156] urging that Vint be retained, and it transpired that Vint would be prepared to stay in Kenya if the pay could be found for him, which it was. The accusation that Vint's proposed transfer was some sort of plot by the Kenyan or British governments to prevent the continuation of the research into mental capacity seems groundless: the impulse to leave Kenya or at least to gain promotion elsewhere seems to have come largely from Vint himself.[157]

The controversy continued, however, over Gordon's post as Consulting Physician to the Mathari Mental Hospital in Nairobi. In 1937 it was decided that a full-time doctor was needed and because of Gordon's age (he was by this time just over seventy), he was not appointed. The underlying reason for this decision was Gordon's obstructive behaviour towards the attempts to reform Kenya's antiquated lunacy legislation. Gordon saw this as a dismissal and also attributed it to official hostility to his thinking on African intelligence. Graham-Little again took the issue up in Britain, making a speech during a Colonial Office debate on the subject in June 1937 and writing to Ormsby-Gore, arguing that Gordon had been badly treated.[158] Graham-Little asserted in the journal, *Empire Review*: 'By this step the Colonial Office has destroyed all prospect of continuing this research.'[159] A resolution was also passed by the Kenyan branch of the BMA expressing disapproval at the termination and the manner of the termination of Gordon's contract and urging that his services be retained.[160] In the case of Gordon, however, the British government was not prepared to be swayed and Gordon was not taken back. Despite Gordon's outrage, the support of the local branches of the BMA, and his recourse to solicitors, the Colonial Office stood its ground. Hand in hand with this dismissal of Gordon from Mathari went an irredeemable rejection of his plans concerning an inquiry into African intelligence. Flood finally wrote of the Kenyan doctors involved in the mental backwardness campaign: 'there is no idiocy and no depth of false reasoning and distortion of phrases to which they would not sink and I really think the best thing to do is to let them alone for a bit and see what else they say.'[161]

The final blow to the Kenyan eugenics campaign came with the publication of Hailey's *African Survey*. As described earlier, the Kenyan eugenicists attempted to harness the feeling that there was a need for research into the state of Africa to gain support for their work on African backwardness and mental capacity. Gordon had hoped that his research might dovetail with Hailey's and he was eager for Hailey to hear him lecture on the subject of backwardness.[162] Those involved with *An African Survey*, however, did not want to become linked to the Kenyan research. As Professor Coupland (Professor of Colonial History at All Souls College, Oxford) put it:

> Miss Matheson has shown me in confidence a copy of Dr Gordon's research-scheme. I think you will agree with me in regarding the sociological side of his scheme amateurish and inadequate. Such a scientific or physicological enquiry as he recommends could, of course, be undertaken without overlapping with my Oxford schemes.[163]

When Hailey was in Kenya in 1936, he did hear Gordon present his paper on 'An Inquiry into the Correlation of Civilization and Mental Disorder in the Kenya Native' at the Kenya Branch of the BMA.[164] He referred to Gordon's work in *An African Survey*, but dismissed its value:

> There would seem, in consequence, little advantage to be derived from the suggested inquiry into the physical characteristics of the African brain as a basis of conclusions regarding his mental capacity. Certainly research of this type would not produce results of the social and political importance which some have expected from it. As regards the proposed inquiry of a more general nature into the causes and the assumed 'backwardness' of the Kenya tribes, while research into the physical characteristics or the psychology of the African is to be welcomed, such inquiry tends to lose its value as soon as it ceases to be objective, and it would not be reasonable to suggest support for an investigation based on assumptions such as those which characterize the theory of *bradyphysis*.[165]

'Bradyphysis' was a term used by Gordon to describe the backwardness of Africans in Kenya; it meant lack of capacity for self-development. Both Gordon and his son-in-law, Humphrey Slade (the barrister and future speaker in the first Independent Kenyan Parliament), wrote letters to *The Times* in outrage at the treatment given by *An African Survey*. Gordon's rather confused letter decried 'This new assumption that African backwardness is after all not a fact'.[166] Slade argued that to deny the concept of 'bradyphysis' potentially made the concept of colonial trusteeship hypocritical.[167] Neither letter was published; copies were sent to the British Eugenics Society and the *Eugenics Review*, which was willing to publish Gordon's letter, but the journal's circulation was considered too small for it to be useful.[168] The response of *An African Survey* brought an end to the Kenyan eugenicists' campaign for support in the metropole: Gordon and Slade's letters of defence failed to interest the press and it was decided in the Colonial Office that 'Hailey's conclusions seem to be grounds to "put by" this issue finally'.[169]

Notes

1 Letter to Hemming, at the EAC, 21 November 1933, CMAC, SA/EUG/C.129.
2 Dawson of Penn, Lord Horder, H. Rolleston, A. Keith and E. Smith to *The Times*, 25 November 1933, p. 8. A copy of this letter was also printed in the *EAMJ*, 10/9 (1933), p. 282.
3 Dawson of Penn etc. to *The Times*, 25 November 1933.
4 Draft of proposed letter to the CO and the Economic Advisory Committee, November 1933, CMAC, SA/EUG/C.129.
5 Gordon to Blacker, 17 November 1933, CMAC, SA/EUG/C.129.
6 Letter to Hemming, at the EAC, 21 November 1933, CMAC, SA/EUG/C.129.
7 D. Kevles, *In the Name of Eugenics: Genetics and the Uses of Human Heredity* (Cambridge, 1995), p. 166.
8 G. R. Searle, 'Eugenics and Politics in Britain in the 1930s', *Annals of Science*, 36 (1979), p. 159.
9 Kevles, *In the Name of Eugenics*, pp. 164–75.
10 J. Huxley, *Africa View* (London, 1931), p. 379.
11 J. Huxley, 'Biology and the Biological Approach to Native Education in East Africa' (April, 1930), NA, CO 879/123. See also 'Memorandum Prepared by Julian Huxley and Dr W. K. Spencer for the Advisory Committee on Education in Tropical Africa', CO 822/15.
12 J. Huxley, *Memories*, Volume 1 (London, 1970).
13 Grant to E. Huxley, 24 October 1934, RH, MSS.Afr.s.2154/1/2.
14 Blacker to Gordon, 9 September 1930, CMAC, SA/EUG/C.129.
15 Blacker to Gordon, 29 October 1930, CMAC, SA/EUG/C.129.
16 Gordon to Blacker, 5 October 1930, CMAC, SA/EUG/C.129.
17 H. L. Gordon, 'Report of a Survey of the Inmates of Kabete Reformatory', p. 19, CMAC, SA/EUG/C.129.
18 Ibid.
19 Secretary to Gordon, 21 January 1931, CMAC, SA/EUG/C.129.
20 Blacker to Gordon, 27 March 1933, CMAC, SA/EUG/C.129.
21 Blacker to J. Huxley, 13 November 1933, CMAC, SA/EUG/C.129.
22 H. L. Gordon, 'Amentia in the East African', *Eugenics Review*, 25/4 (1934), pp. 223–35.
23 Blacker? to Horder, 13 November 1933, CMAC, SA/EUG/C.129.
24 S. Dubow, *Scientific Racism in Modern South Africa* (Cambridge, 1995), pp. 85–6.
25 Gordon to Blacker, undated, *c.* November 1933, CMAC, SA/EUG/C.129.

26 Kevles, *In the Name of Eugenics*, pp. 207–8.
27 Blacker to Gordon, 30 December 1935, CMAC, SA/EUG/C.130.
28 See E. Barkan, *The Retreat of Scientific Racism: Changing Concepts of Race in Britain and the United States between the World Wars* (Cambridge, 1992), p. 266.
29 Huxley to *The Times*, 18 December 1933, p. 8.
30 P. Mockerie, *An African Speaks for His People* (London, 1934), p. 59.
31 Huxley's Foreword to Mockerie, *An African Speaks for His People*, p. 7.
32 Gordon to Blacker, 28 July 1938, CMAC, SA/EUG/C.130.
33 A. F. Tredgold, *A Text-Book of Mental Deficiency (Amentia)* (London, 1937), p. 124.
34 F. W. Vint, 'A Preliminary Note on the Cell Content of the Prefrontal Cortex of the East African Native', *EAMJ*, 9/2 (1932), pp. 30–55.
35 R. J. A. Berry, 'Mental Deficiency in England: An Analysis of the Mental, Physical and Mental Characteristics of a group of 162 Adult Feeble-Minded Women' in R. J. A. Berry (ed.), *Stoke Park Monographs on Mental Deficiency and Other Problems of the Human Brain and Mind* (London, 1933), pp. 65–80, 80.
36 H. L. Gordon, 'The Case of Charles William Ross, Hanged for Murder in Kenya Colony' in Berry (ed.), *Stoke Park Monographs on Mental Deficiency*, p. 237.
37 Blacker to Gordon, 11 April 1934, CMAC, SA/EUG/C.130.
38 'Dr H. L. Gordon', *Eugenics Review*, 39/3 (1947), p. 108.
39 Ibid.
40 Gordon to *The Times*, 8 December 1933, p. 15
41 Leakey to *The Times*, 13 December 1933, p. 15
42 Burt to *The Times*, 15 December 1933, p. 10.
43 Kevles, *In the Name of Eugenics*, p. 84.
44 Ibid., p. 140.
45 Burt to *The Times*, 15 December 1933, p. 10.
46 Ibid., p. 10.
47 Huxley to *The Times*, 18 December 1933, p. 8.
48 Haldane to *The Times*, 19 December 1933, p. 8. The issue of brain size was also picked up by other correspondents: Sunderland to *The Times*, 15 December 1933, p. 10; Pullar to *The Times*, 16 December 1933, p. 15.
49 Gordon to Blacker, 28 July 1938, CMAC, SA/EUG/C.130.
50 Dubow, *Scientific Racism in Modern South Africa*, p. 191.
51 Gordon to Grant, 6 October 1935, RH, MSS.Afr.s.2154/1/3/.
52 Gordon to *The Times*, 22 January 1934, p. 8.
53 Quoted in Kevles, *In the Name of Eugenics*, p. 133.
54 Levo to *The Times*, 19 December 1933, p. 8.
55 Keane to *The Times*, 27 December 1933, p. 15
56 Drummond Shiels to *The Times*, 28 December 1933, p. 6.
57 Gilks to *The Times*, 30 December 1933, p. 6.
58 Gordon to *The Times*, 22 January 1934, p. 8.
59 Graham-Little to *The Times*, 28 August 1934, p. 6.
60 Elliot Smith to *The Times*, 31 August 1934, p. 6, and Dawson of Penn etc. to *The Times*, 3 September 1934, p. 8.
61 Cullen Young to *The Times*, 6 September 1934, p. 8.
62 Editorial, *The Times*, 9 October 1934, p. 15.
63 'The Brain of the East African Native', *EAS*, 30 April 1932, p. 45.
64 Cited in 'The Brain of the East African Native', *EAS*, 30 April 1932, p. 45.
65 'The Mind and Brain of the Kenyan Native', *Lancet*, 25 November 1933, p. 1221.
66 'The Brain of the African Native', *Lancet*, 17 February 1934, p. 357.
67 BMA Council, 1931–32, p. 261. BMA archives.
68 J. H. Sequeira, 'The Brain of the East African Native', *BMJ*, I (1932), p. 581.
69 Editorial, *BMJ*, 30 March 1932, pp. 812–13.
70 N. Leys to *BMJ*, 30 March 1932, p. 826.
71 Editorial, 'The Mind and Brain of the Kenya Native', *BMJ*, 18 November 1933, pp. 931–2.
72 Searle, 'Eugenics and Politics in Britain in the 1930s'.

73 H. L. Gordon, 'The Mental Capacity of the African', *Journal of the African Society*, 33/132 (1934), p. 227

74 *Report of the Commission on Closer Union of the Dependencies in Eastern and Central Africa (Hilton Young Report)*, Cmd. 3234 (London, 1929), and *Memorandum on Native Policy*, Cmd. 3573 (London, 1930).

75 *Report of the Joint Select Committee on Closer Union in East Africa*, H.C. 156 (1931).

76 R. G. Gregory, *Sidney Webb and East Africa: Labour's Experiment with the Doctrine of Native Paramountcy* (Berkeley, 1962) and D. Wylie, 'Confrontation over Kenya: The Colonial Office and Its Critics, 1918–1940', *Journal of African History*, 18/3 (1977), pp. 427–47.

77 A. D. Roberts, 'The Gold Boom of the 1930s in Eastern Africa', *African Affairs*, 85/341 (1986), pp. 545–62.

78 G. Bennett, 'Settlers and Politics in Kenya up to 1945' in V. Harlow and E. M. Chilver (eds), *Oxford History of East Africa* (Oxford, 1965).

79 B. Berman, *Control and Crisis in Colonial Kenya: The Dialectic of Domination* (London, 1990), pp. 197–8.

80 P. Hetherington, *British Paternalism and Africa, 1920–1940* (London, 1978), p. 90.

81 Ibid.

82 R. D. Pearce, *The Turning Point in Africa: British Colonial Policy, 1938–1948* (London, 1982).

83 Ibid., p. 3.

84 Gregory, *Sidney Webb and East Africa*, p. 27.

85 Wylie, 'Confrontation over Kenya', p. 429.

86 NA, CO 533/648.

87 Strachey, 'Kenya Native Welfare Organisation', 24 August 1926, NA, CO 533/648.

88 Oldham to Grigg, 10 January 1927, Bodleian, MSS 1002, General Correspondence, 1925–1930.

89 For its valuable account of *An African Survey*, see H. Tilley, 'Africa as a Living Laboratory: The Colonial Research Survey and the British Colonial Empire', PhD thesis, University of Oxford, 2001.

90 'The Mind of the Native', *EAS*, 7 October 1926, pp. 7–9. Continued in *EAS*, 8 October 1926.

91 Ibid., p. 7.

92 Oldham to Ormsby-Gore, 'Research into Native Welfare in East Africa', 10 September 1926, NA, CO 533/648.

93 Ibid.

94 *Education Policy in British Tropical African Dependencies*, Cmd. 2374 (1925), p. 4.

95 Oldham to Ormsby-Gore, 11 February 1929, SOAS, IMC/CBMS/219 (Archives of the International Missionary Council and the Conference of British Missionary Societies, held at SOAS).

96 C. Parkinson, *The Colonial Office from Within, 1909–1945* (London, 1947), p. 58.

97 Mayhew, 'A Note on Psychological Tests of Educable Capacity & on the Possibility of Their Use in the Colonies,' p. 2, SOAS, IMC/CBMS/223.

98 Gordon to *The Times*, 8 December 1933, p. 15.

99 Mayhew to Oldham, 8 December 1933, SOAS, IMC/CBMS/219.

100 Mayhew to Oldham, 14 December 1933, SOAS, IMC/CBMS/219.

101 Matheson to Oldham, 14 December 1933, RH, MSS.Afr.s.1829/1/3.

102 Paterson to Major Wells, 17 January 1934, NA, CO 822/61/14.

103 Colonial Development Fund. Answers to Questionnaire, p. 1, KNA, BY/26/7.

104 Munro to Stanton, 4 May 1932, NA, CO 822/47/5.

105 Ibid.

106 Memorandum by Munro, 4 May 1932, NA, CO 822/47/5.

107 Sequeira to Gilks, 10 May 1932, KNA, BY/26//.

108 Gilks to Stanton, 28 May 1932, KNA, BY/26/7. See also NA, CO 822/47/5.

109 Ibid.

110 Ibid.

111 Ibid.

112 Quoted in a letter from O'Brien to Gilks, 31 August 1932, NA, CO 822/47/5.
113 O'Brien to Gilks, 31 August 1932, NA, CO 822/47/5.
114 Note recording telephone conversation with Oldham, 29 September 1932, NA, CO 822/47/5.
115 Gilks to Stanton, 9 December 1932, NA, CO 822/47/5.
116 Gilks to Stanton, 24 September 1932, NA, CO 822/47/5.
117 Extract from Minutes of a meeting of the Colonial Advisory Medical Committee, 4 October 1932, NA, CO 822/47/5.
118 Gilks to Stanton, 9 December 1932, NA, CO 822/47/5.
119 Comment by Stanton, 12 August 1932, NA, CO 822/47/5.
120 Report of the Proceedings of the AGM of the Empire Parliamentary Association (UK Branch), 24 July 1934, HLRO (House of Lords Record Office), CPA/13.
121 Extract from Official Report, 29 November 1933, CMAC, SA/EUG/C.129.
122 Memorandum by Flood, 9 November 1936, NA, CO 822/72/8.
123 Memorandum by Hemming, 19 October 1933, NA, CO 822/55/1.
124 Moore to Flood, 3 November 1933, NA, CO 822/55/1.
125 Note written by Hemming recording a conversation between Hemming and Freeston, 16 November 1933, NA, CO 822/55/1.
126 Flood to Moore, 8 December 1933, NA, CO 822/55/1.
127 Quoted by Berman, *Control and Crisis in Colonial Kenya*, p. 180.
128 Comment by Stanton, 12 July 1932, NA, CO 822/47/5.
129 Comment by Stanton, 18 October 1933, NA, CO 822/55/1.
130 Flood to Moore, 8 December 1933, NA, CO 822/55/1.
131 Gordon to *The Times*, 8 December 1933, p. 15.
132 Flood to Moore, 8 December 1933, NA CO 822/55/1.
133 Memorandum by Flood, 9 November 1936, NA, CO 822/72/8.
134 Note by Flood, 8 March 1934, NA, CO 822/55/1.
135 Memorandum from Paterson to Major Wells, 17 January 1934, NA, CO 822/61/14.
136 Dispatch from the Governor (Byrne) to Cunliffe-Lister, 5 July 1934, NA, CO 822/61/14.
137 Freeston to Wade, 4 September 1934, NA, CO 822/61/14.
138 CO to Scott, Mitchell, Hall, Dundas and Lawrence, 11 October 1934, NA, CO 822/61/14.
139 Extract from Minutes of a Meeting of the Colonial Advisory and Medical Committee, 16 October 1934, NA, CO 822/61/14.
140 Eleanor Rathbone to Cunliffe-Lister, 7 September 1934, NA, CO 822/61/14; Collinson (Hon. Political Secretary of British Commonwealth League) to Cunliffe-Lister, 21 September 1934, NA, CO 822/61/14; Grobel (Gen. Sec. of National Council of Women of Great Britain), 23 November 1934, NA, CO 822/61/14.
141 Comment by Flood, 18 October 1934, NA, CO 822/61/14.
142 Hemming to Freeston, 16 November 1934, NA, CO 822/61/14.
143 Tilley, "Africa as a Living Laboratory", p. 227.
144 Ibid.
145 Comment by Freeston, 10 October 1934, NA, CO 822/61/14.
146 'African Research Survey', *Bulletin* of The Women's Freedom League, 26 June 1936, p. 5, NA, CO 822/72/7.
147 Graham-Little to *BMJ*, 12 September 1936, pp. 560–1.
148 Memorandum by Flood, 9 November 1936, NA, CO 822/72/7.
149 Comment by Flood, 26 October 1936, NA, CO 822/74/10.
150 '"Medical Crisis" in East Africa', *EAS*, 11 November 1936, p. 23.
151 Memorandum by Flood, 9 November 1936, NA, CO 822/72/7.
152 '"Medical Crisis" in East Africa', p. 23.
153 Ibid., p. 23, see also Graham-Little's letter to *BMJ*, 31 October 1936, p. 892.
154 Paterson to Stanton, 22 December 1936, NA, CO 822/78/19.
155 Petition to Sir Joseph Byrne, 29 September 1936, NA, CO 822/74/10.
156 Telegram from Governor Byrne to Ormsby-Gore, 7 October 1936, NA, CO 822/74/10.
157 Comment by Flood, 23 March 1937, NA, CO 822/78/19.

158 Extract from a speech made by Graham-Little, 2 June 1937, NA, CO 822/78/19; letter from Graham-Little to Ormsby-Gore, 2 May 1937, NA, CO 822/78/19.
159 E. Graham-Little, 'British Empire and Backward Races', *Empire Review and Magazine*, 66/442 (1937), pp. 269–75, 275.
160 'A Grave Discourtesy', *East Africa*, 1 April 1937, NA, CO 822/78/19.
161 Flood to McLean, 3 May 1937, NA, CO 822/78/19.
162 Grant to Huxley, 11 July 1934, RH, MSS.Afr.s.2154, File 2
163 Coupland to Hailey, 13 March 1936, RH, MSS.Afr.s.1829/1/3.
164 H. L. Gordon, 'An Inquiry into the Correlation of Civilization and Mental Disorder in the Kenya Native', *EAMJ*, 12/11 (1936), pp. 327–35.
165 M. Hailey, *An African Survey* (London, 1938), p. 38.
166 Unpublished letter from Gordon to *The Times* (undated but enclosed with a letter to Blacker dated 15 January 1939), CMAC, SA/EUG/C.130.
167 Unpublished letter from Slade to *The Times*, 24 November 1938, CMAC, SA/EUG/C.130.
168 Blacker to Major Granville Edge, 23 January 1939, CMAC, SA/EUG/C.130.
169 Comment by F. J. Pedler, 18 November 1938, NA, CO 822/85/11.

CHAPTER 5

Settler attitudes to eugenics and race

The eugenics movement that emerged in Kenya in the early to mid-1930s both chimed and at times subtly clashed with settler prejudices and preoccupations. The movement was born out of British eugenics – a eugenic hybrid was created, which used the same intellectual framework and attracted a similar audience to that of British eugenics, but which was also distinctively motivated by a combination of racial hostility, paternalism and settler political ambitions. This chapter will shed light on eugenics in Kenya through an examination of the social composition and interests of the eugenicists and by contextualising their ideas, showing how eugenic thought shaped itself to the local social and political situation.

The Kenya Society for the Study
of Race Improvement (KSSRI)

A Society for the Study of Race Improvement has been formed in Kenya. The inaugural meeting was attended by leading professional and business men and some 40 members were enrolled.

Broadly, the objects of the Society are the diffusion of suitable knowledge in Kenya on race improvement and the endeavour on the part of members to pave the way for the firmer footing of generations to come.

In the course of addresses it was stressed that whilst much attention is devoted to the careful breeding of animals, too little thought has been spent in the past on improving the physical and mental equipment of human beings.[1]

The interest that eugenics aroused in Kenya was enough to lead to the establishment of the KSSRI in 1933. Dr Gordon had already been regularly making public lectures on the subject of eugenics and it was through the medium of these popular lectures that support for eugenics in Kenya was initially elicited. Early in 1933 Gordon performed a lecture on eugenics under the auspices of the Natural History Society which was so successful that a repeat was requested, this time in the Theatre Royal in Nairobi to accommodate a larger audience. This was held in March 1933 and it was at this lecture that Mrs Shaw suggested a race improvement society be formed.[2] This initial lecture and the subsequent formation of the KSSRI received extensive coverage from the main settler newspaper, the *East African Standard* (the *EAS*).[3] This may have partly been because the man who founded the newspaper, its one-time Editor and current Managing Director, was Mr R. F. Mayer, the Chairman of the KSSRI.

Gordon wrote to the Eugenics Society in Britain describing the level of interest shown in the colony:

> I have just given two lectures on eugenics and there is a hot desire to start a Race Study and Improvement Society. Does the Eugenics Society gives its blessing to such efforts in any way?
>
> A very few pamphlets I had were simply fought for after my first lecture.[4]

As well as gaining the support of the Eugenics Society, letters offering help were also received from the British groups, the National Council of Mental Hygiene and the Child Guidance Council. South African literature on the question of poor whites was also circulated among Kenyan eugenicists.[5] The eugenics movement in Kenya developed its own characteristics, particularly in the area of research on African intelligence, but Gordon and his colleagues also saw it as an imperial movement, with international implications. This imperial conception meant that the KSSRI frequently had its eye to the metropole for ideological and financial clout, although it was only in respect of the former that the Society succeeded in benefiting. This touches on a significant aspect of the tone of the KSSRI, which affected its relationship with the settler community as a whole. The members of the KSSRI were eager to establish the Society as a part of an international intellectual elite. There was great emphasis on eugenics as a worldwide movement and on the status of its Kenyan members, as 'great leaders' in the areas of 'Church, Medicine, Law and Science'.[6]

The membership of the KSSRI in 1933 seems to have amounted to about sixty.[7] In 1932–33 membership of the British Eugenics Society was

---

The author would like to draw attention to an error that she has discovered on pp. 118, 161 and 169 in the first edition of this book, which has been corrected in the paperback reprint. Mrs Shaw, who acted as secretary for the Kenya Society for the Study of Race Improvement, was not the Mary Shaw (Dr M. M. Shaw) who worked as a medical officer in charge of Child Welfare in Kenya in the early 1930s. There is no evidence that Dr Mary Michael Shaw was involved in the KSSRI.

at 768.[8] Given that the KSSRI only permitted European members, it attracted a much greater number in relation to overall population than the British Eugenics Society (the European population in Kenya was 16,812 in 1931).[9] What is more, this did not require much effort; the KSSRI was not particularly concerned with attracting a larger membership, as Eleanor Grant put it: 'It is a lovely Cttee – all the doctors don't hold w. Cttees much and chafe and fidget till they get a chance of letting themselves go on their subjects, & then they are divine. The Cttee doesn't want any active propaganda done at the present, which is lovely, and no money collecting.'[10] Membership numbers do not tell the whole story. As Searle has argued in relation to British eugenics, the emphasis was not on mass recruitment, but on attracting influential and in particular academically esteemed members, the success of this policy in Britain in the 1930s being demonstrated by the large number of professors among the membership.[11] The lack of enthusiasm for propagandising indicated in the above quotation was possibly attributable to a similar ethos among Kenyan eugenicists, a desire to restrict the actual membership of the organisation to a small, select group. Judging by the descriptions of sizes of audiences at public meetings, which permitted non-members, there was certainly a large amount of interest in eugenics among European settlers more widely. Eleanor Grant gives the frustratingly vague description of there being '100s of people' at the first meeting of the KSSRI.[12] There was certainly enough interest in 'Race Improvement' to lead to the foundation of another branch of the KSSRI at Nakuru in November 1933: 'On Friday evening I took the Chair at Nakuru's 1st Race Improvement Meeting. Sequeira read his paper on Parent Problems and Problem Parents. There was an extraordinarily good audience of about 40, all sorts of people you wouldn't expect, like G. J. Dawson and wife, all the doctors in the crowds.'[13] Grant also wrote of people 'flocking in to the RI [Race Improvement] Society'.[14]

The letters written by the Honourable Eleanor Grant to her daughter, the writer Elspeth Huxley, provide a useful account of the social composition and attitudes that characterised the eugenics movement in Kenya. Grant's involvement in the KSSRI is itself interesting. She was of the settler social elite; her values, in particular her devotion to the creation of a rural, agricultural idyll in the colony, reflected the anti-urban inclinations of settler society. Along with this conservatism, Grant also displayed a rather modern interest and faith in the progressive nature of science, particularly biology; her letters convey a typically eugenic espousal of a materialist, biological pragmatism that she placed in opposition to Victorian squeamishness and sentimentalism. Beneath the gossipy, *faux* naïve tone of her letters, Grant captured the intellectual aspirations that motivated settler eugenics. Elspeth Huxley, despite not living in Kenya as an adult, became the settlers' literary voice, writing a biography of Lord Delamere and various articles and stories about the colony in defence of the settler perspective.[15] In 1980 Huxley published a memoir of her mother which included her letters; significantly, she excised all the comments on eugenics and the KSSRI that Grant had frequently made in the original letters.[16]

Dr Gordon described those interested in his lectures as 'mostly British of the intelligent middle-upper class, including civil servants.'[17] The *EAS* reported that the inaugural meeting, held at Nairobi's New Stanley Hotel, was attended by 'leading professional and businessmen'.[18] Grant's description of the first KSSRI meeting also gives an indication of the social composition of the Society; its members, or the 'race improvers', as Grant called them, tended to move in the circles of Kenya's social and administrative elite. Mrs Moore was on the committee of the KSSRI; her husband, Sir Henry Monck-Mason Moore, was Colonial Secretary in the early 1930s, which involved officiating as Acting Governor for five months in 1933. He returned to Kenya as Governor in 1940, after leaving to become Governor of Sierra Leone in 1934. Lady Shaw was involved with the KSSRI; she was married to Sir Robert Shaw, who was Acting European Elected Member for Ukamba in the Legislative Council in 1933 and fully Elected Member in 1935. Also involved in the KSSRI were Captain H. F. Ward and his wife. Captain Ward was European Elected Member for Nairobi North. Mrs Hodge, who was on the committee of the KSSRI, was married to the Honourable S. O. V. Hodge who became Provincial Commissioner for the Rift Valley in 1938.

The Chairman of the Society was R. F. Mayer, OBE, an influential figure in the European settler community owing to his control of the *EAS*. In fact, it seems that the approval of Mayer was of crucial importance in achieving the public arena necessary for such groups; the Natural History Society failed to elicit much public interest after its organiser, van Someren, had a row with Mayer, who withdrew publicity.[19] Dr van Someren was Vice-Chairman of the KSSRI when it was first founded, and was a well-established figure in the medical sphere in Kenya. He had been the Dental Surgeon in the Medical Services in the 1920s, but in 1907 he was also running the precursor to the mental asylum in Kenya, a camp for lunatics, and was then in charge of the actual asylum until 1914.[20] Dr Sequeira was the other Vice-Chairman of the KSSRI. Another influential member on the KSSRI committee was the Reverend R. V. Wright, Dean of Nairobi Cathedral.

Grant also recorded that Hugh La Fontaine, who was at that time the Acting Chief Native Commissioner, had a prominent role at a race improvement meeting 'in support' of Gordon.[21] The position of Chief Native Commissioner, albeit acting, was a senior one in Kenya. In the

1930s La Fontaine was involved in formulating policy on issues such as African education and delinquency, writing an influential report in 1933 advising on the introduction of Borstal methods to Kabete Reformatory[22] and producing a report on the Independent Schools Movement in 1936.[23] La Fontaine was one of the more forward-thinking and progressive colonial officials in Kenya. Askwith, who later became Commissioner for Community Development, remembered La Fontaine as being one of the few Europeans in Kenya who were outspoken in their dislike of racial discrimination and segregation.[24] La Fontaine was an active founder member of the United Kenya Club; founded after World War Two and still existing today, it was a club that aimed to promote racial co-operation. Its membership was open to all races and it provided a socially equal meeting place.[25]

The rules for the KSSRI stated that membership was only open to Europeans, despite the fact that members of the Indian community attended the public lectures, as Gordon put it: 'Possibly 10 per cent would be Indian, equally intelligent [as his European audience] but recklessly prolific in procreation.'[26] There is no other mention of Indians being involved in the Kenyan eugenics movement; clearly, this was largely because they were excluded from involvement at a higher level than as members of the audience in the public meetings.

The topics discussed under the auspices of the society were a blend of traditional eugenic interests and local Kenyan concerns. The emphasis was not just on race, but on the broad social concerns that were traditionally encompassed by eugenics in Britain, such as 'Infant welfare, child guidance, cruelty to children, birth control, sex instruction, sterilization of the unfit, mental hygiene, mental deficiency'.[27] One of the first discussions held by the Society was on 'The World Population Problem'. It was opened by Mrs Grant and Mrs Shaw, and Paterson and Sequeira were among those who took part in the debate.

The KSSRI was emphatic about the value of eugenic theories in Kenya:

> It was realised that very few people in the Colony had accurate knowledge on this important subject, and that a society whose object was to supply this information was urgently needed in a community progressing so rapidly.
>
> While it was recognised that a study of heredity and eugenics must be an important part of the Society's programme, various speakers pointed out that race-improvement involved also a careful investigation of the problems of environment, some of which are already urgent in Kenya.[28]

There was a strong emphasis on children in this outline of interests: this reflected the preoccupations of settler culture with the future physical and political potency of the white presence.

The KSSRI tended to be disparaging about missionaries and their capacity to grasp the difficulties facing Kenya, an attitude that seems to have been shared by much of the settler community, particularly after the 1928–30 female circumcision crisis.[29] It is striking that there is no evidence of any missionary involvement in the KSSRI or of medical missionaries supporting the theories of Gordon and Vint. There was no mention of the Kenyan research in *The Mission Hospital*, the Church Missionary Society's journal of its medical activities, despite the fact that there were two CMS hospitals in Kenya at the time whose doctors wrote reports for the journal.[30] In a letter to Blacker of the British Eugenics Society, written at the beginning of the controversy over African mental capacity, Gordon, while emphasising his own objectivity, described how his stance had already alienated the missions and earned the support of those hostile to African development:

> Here the black v. white controversy obscures everything and creates an atmosphere inimical to progress. One side regards the native without hope as inherently deficient; the other is prepared to raise him by 'education' to our level and a share in government, in a 'generation or two'. I venture to deprecate prejudice and to put (and stick to) the simple question: why not see what science has to say on this question of the inherent capacity of the native? More pious approval poured in from the first party. The second party, which includes the powerful missionary organizations, stamped me as an ally of the other fellows – which makes me feel I bungled.[31]

The Dean of Nairobi, the Reverend Wright, was an active member of the movement, but he was not a missionary. The incompatibility of missions and eugenics in Kenya has various underlying causes, all of which contribute to our understanding of the role of eugenics in Kenya. For a start, eugenics was never accepted by the Roman Catholic Church; in December 1930 Pope Pius XI condemned eugenics in his encyclical *Casti Connubii*, in which he declared that the spirit was paramount over the body and the equality of human souls regardless of material defect. Other missionaries, although not doctrinally opposed to eugenics in the same way, were likely to be suspicious of eugenics because of a longstanding conflict between traditional Christianity and the modern, atheist, Darwinian ideological roots of eugenics. The division between these traditions was compounded in Kenya by the question of native interests; the education and development being encouraged by missionaries was of precisely the kind that the Kenyan eugenicists were warning might be unsuitable and even dangerous when applied to Africans

A letter written to the *EAS* criticising the KSSRI raised complaints about the agenda of the Society that chime with typical Christian doubts about eugenics and its secular modernity at the expense of traditional

Christianity. The letter raised concern at the emphasis of the KSSRI on the material rather than the spiritual aspect of mental wellbeing and the suspicion that eugenics was connected with an undermining of the Christian sexual order:

> It may be remarked that at the above meeting the Churches had no formal representative and as far as we know no Church authority in Kenya has approved of the Society.
>
> It is also known that the Society has *not* the *unanimous* approval of the Medical Profession; and it is a little premature and possibly not *fair* to launch the movement as if it had the blessing and encouragement of 'real religion and real Science' ...
> (a) Does the Society consider the increasing number of divorces an aid to Race Improvement?
> (b) Does the Society consider the increasing number of sex novels, sex plays and sex cinema shows – with their lurid sexual street advertisements – an aid to Race Improvement?
> (c) Would the originators in Kenya of the Society kindly inform us what they mean by the vague terms, 'unfit' 'heredity' 'eugenic' and 'conscience'?
> (d) Does the Society ignore spiritual forces in Race Improvement and, if not, how does it propose to include these forces in its activities?[32]

This letter was not necessarily written by a missionary – it was only signed 'Inquirer' – but it does draw attention to how alarming and incompatible Kenyan eugenics were for the missionary agenda. An indication of settler attitudes to the missionary stance on race and intelligence is given in a letter written to the *EAS* signed under the pseudonym 'George William', in which Scott's dissenting note on African educable capacity was dismissed as representing 'the old "missionary" case'.[33]

The KSSRI seems to have petered out in about 1935 as it lost some of its most important members:

> I don't expect to go soon to N'bi, as am resigning from the Race Improvers Cttee. I fear the poor Race Improvers may be doomed to extinction, or at least a prolonged period of dormancy, as Mayer was chairman & is dead, Mrs M. M. Moore gone ... Cicely Shaw is at Mombasa, Molly Hodge goes on leave in May & so on.[34]

The impetus behind the KSSRI came from doctors like Gordon and Sequeira, who had changed the focus of their campaign by this time. Their interest lay in the metropole and the acquisition of support from the imperial government.

Eugenics and settler politics

In this section, the connections between the eugenic agenda in Kenya and key political issues that confronted and challenged European settler dominance and authority will be examined. Members of the KSSRI themselves denied political bias, as Mayer grandly stated:

> this world-wide movement in the cause of humanity has no political object, its principles are equally in harmony with the ideals of all political parties and appeal to all who have the interests of the human race at heart no matter what class or party they belong to.[35]

Although there was undoubtedly a range of social and political motivations behind involvement with the KSSRI, it will be argued here that many of the assertions of political objectivity made by the race improvers were somewhat disingenuous when they are considered within the contemporary settler political environment.

In the 1950s, in the context of the racial hostility aroused by Mau Mau and also recent memories of Nazism, the racist meaning and purpose of Kenyan eugenics could not be ignored. But in the 1930s, when the Kenyan eugenicists were presenting their ideas, their denials of political bias did not appear so weak. As Redley has pointed out, the self-regulating consensus on race was so strong that it was not a political issue.[36] We need to distinguish where Kenyan eugenics was reflecting, and fortifying, a more diffusely racist consensus, and where it was appropriating science to defend settler racial interests in issues of particular political moment.

Although, functionally, the opinions of the Kenyan eugenicists can be clearly seen as complicit with broader settler interests, this was not consistently perceived to be the case at the time. The very fact that the Kenyan eugenicists devoted so much attention to native backwardness was questioned at the inaugural meeting of the KSSRI. The issue was raised by a settler, Mr Payne, who suggested of the KSSRI that:

> If it confined itself at the outset to Europeans it would do an immense amount of good.
>
> He urges that whites should be taught hygiene and development rather than run the risk of trying to take the black races across a period of a thousand or 1,500 years at one step to the present stage of European development and teach him what was after all, a very dangerous subject.
>
> The Chairman pointed out that the speaker had touched on a very dangerous subject, and reminded him of the title of the Society – the Kenya Society for the Study of Race Improvement. The white population, added Mr Mayer, were the trustees of the native population, and he personally could never assent to the views advanced by Mr Payne, to leave the native out of consideration by the Society. The native races in this country really wanted improving more than the European race.[37]

The pursuit of African development by the KSSRI was a matter of some concern to European settlers who were jealously hostile to the promotion of African interests. An interesting facet of Kenyan settler society, which is not conveyed, for example, in Kennedy's *Islands of White*, is not how preoccupied settlers were with Africans but in fact how uninformed and uninterested they were. The letters and papers of settlers are often striking in their absence of comment and thought on Africans, resulting in an almost universal tone of amused cliché and sometimes paternalistic, sometimes overtly hostile dismissive bemusement, which was expressed in consistently recurring platitudes. As Duder and Youé put it:

> Beyond individual relationships, the settlers often dealt with Africans by simply ignoring them. Contemporary descriptions of Nanyuki township contain no mention whatsoever of African residents in the town ... It was as if they were not there.[38]

The pages of the *EAS* indicate that settlers normally engaged in debates about Africans only when it came to the question of labour or taxes. The interest of the KSSRI in the African brain and psychology, despite its negative assessment of Africans, at one level clashed with the indifference to Africans shown by many settlers, particularly as this absence of interest and engagement in African conditions was an expression of supremacy: the subtext of this indifference to the mental life of Africans was an assertion of the chasm that lay between Africans and Europeans.

Although the Kenyan eugenicists argued that East Africans were incapable of mental development equivalent to European standards, the argument was framed in terms of combating native backwardness, which in the context of the political extremity of settler culture, gave the subject liberal pretensions. The response to Payne's complaint demonstrates this. Gordon admitted that Payne had raised an important question and conceded that it was only natural that the first object of the society should be to improve their own race. In doing so, Gordon argued that they would also be 'indirectly improving the native races, because we are raising our own standards, and they learn far more than is realised by imitation and suggestion'.[39] Gordon also argued that science would 'deprecate the Society not concerning itself with the Asian and native races':

> In thinking of the future of Kenya, we are not thinking of our own selfish predominance. If the native improves, we must improve *pari pasu*, The [sic] gist of the scientific position is this – that science is interested in the African only in order to discover what the causes of African backwardness are, and if it is possible to discover better means of raising the African out of that backwardness than we now employ.[40]

La Fontaine commented that he hoped the society would not restrict its work to Europeans as he believed there were many ways the KSSRI could help with 'the improvement of the natives'. He also suggested that district officers should be kept in touch with what the KSSRI was doing.

The Kenyan eugenicists were on occasion quick to rebut statements by settlers that attempted to co-opt the research on race and intelligence into the cannon of arguments used to defend European supremacy. For example, in 1934, when the question of African representation on the Legislative Council was raised, a Mr Davis wrote to the *EAS*, citing Gordon's research as an argument in his attack on the idea, saying the notion of African enfranchisement:

> ignores the question again raised recently by Dr. Gordon's psychological research on the innate mental capacity of the African as compared with the European stocks, for what Dr. Gordon has expounded on the cranial differences between white and black has been said before. These differences have been shown in the ultimate results of education which in the case of the Native at its peak seems to stop short and diverge at a tangent where in the European it progresses in adult culture. If this theory, based upon research, be accepted then political equality in the future is a vain chimera.[41]

Mayer wrote into the *EAS* to counter this direct politicisation of Gordon's ideas:

> With reference to the remarks made by Mr A. Davis in your issue to-day on the subject of the recent researches on the mental capacity of Africans, I am in a position to state that no such conclusions have been drawn by either Dr Gordon, Dr Vint or any scientific authority acquainted with that work. In fact I would point out that no conclusions of any kind have been drawn.
>
> The matter is *sub judice* and expressions of opinion at this stage can only be injurious to the high interests involved in this important work, which one hopes is only an introduction to greater and wider research.[42]

Mayer's letter was not entirely honest because Gordon and Vint had extrapolated educational and political implications from their work on mental capacity. Gordon in particular had used their findings on that subject as a part of a wider argument about race and intelligence, albeit an argument occasionally qualified by warnings that these were preliminary results which needed confirmation by an extended research project.

The editor of the *EAS* made a comparable response to a letter written by an African, K. Kisosonkole of King's College in Uganda. Kisosonkole, in a very temperate letter, brought up the question of environment, tribal differences, and the case of high-achieving, Western-educated Africans like Dr Aggrey. The editor wrote a note at the end of Kisosonkole's letter that was remarkably disingenuous given the coverage his newspaper had given to the research on race and intelligence and the significance the *EAS* had attributed to the research: 'All readers, especially Africans, should realise that this line of research is only in its infancy and present "conclusions" are merely speculative. Their primary value is encouragement of further study on an organised scale.'[43] Kisosonkole wrote a somewhat more forceful letter in 1934, 'putting down hard figures like this by the scientists has a detrimental effect on the African: his degree of educability will be misjudged and thus his development curtailed and hampered; and what is more his proper place in the world will be denied him longer than otherwise would be necessary'.[44]

The attitude of the Kenyan eugenicists to the defining aspects of settler society described in recent historiography is revealing. In *Islands of White*, Kennedy has described anxieties about the African climate and the African people as central in shaping the values and identity of settler culture in Kenya and Southern Rhodesia.[45] 'Black peril', the fear of the sexual assault of white women perpetrated by black men, was a distinctive rallying point in settler culture, but does not seem to have been directly discussed by the KSSRI; it appears to have been considered symptomatic of the more profound eugenic problem of poor whites. Given the connection between this issue and miscegenation, which Gordon and his confreres warned must be avoided at all costs, this lack of interest is perhaps surprising. Grant, for example, was disapproving of the agitation by the East African Women's League (EAWL) about black peril and related her disapproval to a eugenic-based argument about poor whites:

> I quite agree that we want cheaper & more rural telephones & in some cases, some sort of better policing, but we <u>don't</u> want agitations over rape, & if the E.A.W.L. would work more at the eugenics of the white population, it wd be much more constructive than attempting to contravene justice, & demanding public hangings etc. There seem to be 300 white children getting no education at all, & all the wrong people breeding like rabbits, & of course these sort of people always will get raped etc.[46]

The problem of rape was thus equated with the problem of poor white women and the failings in their behaviour. This followed frequent warnings found in settler literature on how to manage house servants; such warnings were particularly directed to white women whose inappropriate behaviour in the vicinity of male black servants was thought to have diminished the prestige that had once served to protect them.[47] It was felt that the problem increased with a rise in women settlers who were unfamiliar with dealing with servants before their arrival in the colony; in other words, susceptibility towards black peril was linked with class.

Although the social and intellectual elitism of the members of the KSSRI meant that they appeared reluctant to conform to popular settler areas of agitation, the political implications of the theories on race and intelligence were undeniable. As Davis' letter to the *EAS* demonstrates, the research fed into settler racism and became politically potent despite the pleas for restraint made by the likes of Mayer. The literature on South Africa brings a picture of this process more clearly into focus. Both Dubow and Klausen have pointed out the links made between racial purity and strength and motherhood in South African eugenic discourse. Miscegenation became a particular source of concern.[48] For instance, Harold Fantham and Annie Porter conducted research on miscegenation in the 1920s, concluding that racial mixture led to physical abnormalities and mental and moral failings.[49] Hyslop has described the popularisation and politicisation of fears about race purity and miscegenation in the movement initiated by Malan's 'Purified National Party' for the restriction of mixed marriages between white, in particular working-class Afrikaner, women and black men.[50] The reasons for the potency of this image of white womanhood lay, as Hyslop has cogently argued, in gender relations and the need for Afrikaner men to reassert their dominance over Afrikaner women in response to the emergence of a newly independent female Afrikaner proletariat in the 1920s and 1930s. The language of this discourse on miscegenation indicates how eugenic thought could take on a political life outside medical journals and elite organisations to become a part of popular racist rhetoric. Gordon was particularly vehement on the subject of race mixture. He was optimistic on the fate of white settlement in Kenya provided one condition was met: 'there must be no mixture with negroid blood.'[51]

The claims of objectivity made by the race improvers need to be tested against the Kenyan political context in order to understand how eugenics interacted with settler politics, how it complemented the racial interests of the settler political agenda without alienating certain key liberal, 'pro-native' officials. Two areas of particular political significance to settlers will be examined. The first is the validity of Kenya as a 'white man's country', biologically and politically. The second is the question of education, in particular the development of African demands for access to education of a type that was felt to threaten the European monopoly on 'civilisation' which was so vital to the maintenance of European supremacy. Through an analysis of the connections between these issues and the eugenic discourse, the much-vaunted objectivity of many of the Kenyan eugenicists will be disputed.

Self-government, climate and Closer Union

Kennedy has described the widely held conviction of European settlers that solar radiation from tropical sunshine was potentially dangerous to white people, which led to the donning of special hats (sometimes double hats) and pads to protect the spine from exposure. Some people had a red lining sewn into their clothes as an extra precaution. It was felt that even inside the house, some sort of block was needed for fear that the corrugated iron roofing did not provide adequate protection.[52]

One of the most pessimistic medical experts on white settlement in Kenya was Dr Mackinnon.[53] In his presidential address to the Kenya BMA in 1935, Mackinnon considered the question of research into African intelligence in a similar light to the problem of climate and European health in Kenya:

> I am sure we are all very pleased at the success which has so far attended Dr. Gordon and Dr. Vint in their efforts to elucidate some of the peculiarities and abnormalities of the African native brain. We congratulate them both on the great value of their researches and on the interest aroused at home among scientists and members of Parliament. We hope that further stimulus will be given to continue these researches ... If a research is necessary into the mentality of the native in his own natural surroundings I would ask in all seriousness is it not equally necessary and desirable to promote research into the effects of the sun, altitude and other conditions on the well developed brain and constitution of a white race, living under conditions to which they are by nature unfitted?[54]

At one of the early meetings held by the KSSRI, A. Walter, Director of the British East African Meteorological Services, read a paper on 'Climate and White Settlement in the East African Highlands'.[55] Given the importance of ideas about climate within settler culture it is not surprising that this was one of the first topics brought up by the KSSRI.

The reason for this anxiety about climate in Africa lay in its association with racial origins and the identification of those origins with environmental causes. This can be seen in the framing of the problem by Walter in his paper at the KSSRI:

> That climate has been one of the factors determining the differentiation of the human species, or at least those drawn from the same parent stock, will, we think, hardly be contested ... This being the case, it is somewhat surprising that the scientific study of climate should not precede rather than follow the extensive schemes for colonisation. The acclimatisation of the white races in the tropics whether in the Highlands of Kenya, the lowlands of other tropical countries, is a problem fraught with far-reaching consequences.[56]

Walter suggested that problems could arise for Europeans from the effects of the extreme diurnal changes in temperature in the Highlands; the effects on European physiology of adapting to warm days and cold nights, a product of altitude, were unknown. In particular, he warned of the unknown results of the strain of this constant adaptation for children. The consequences of altitude arising from lower oxygen density was warned of, as was the way in which altitude lowered boiling point and potentially lessened the nutritional value of food. Walter also argued that the intensity of the sunlight in Kenya could result in a higher expenditure of nervous energy. Walter did not state that the Kenyan climate was definitely harmful to Europeans, but he listed serious points for concern that needed to be carefully researched. The question of the suitability of Kenya for white settlement was considered of such importance that Walter's lecture to the KSSRI provoked leading articles in the *EAS* and the *EAMJ*.[57] The *EAS* approached the issue from the angle of the eugenic quality of the European stock in Kenya:

> It is true that, taken as a whole, white settlement has been amazingly lucky in the early years of its establishment. It was drawn from good stock, it had an adventurous spirit as an aid to mental alertness and it has maintained until now a very high standard of living.
>
> Yet the fact remains that there has been established in this country a European community composed of people who are unaccustomed to the conditions of their new life. They have been bred from stock which lived in entirely different conditions and who have had to attempt a rapid and revolutionary adaptation without any knowledge of the effects or of the steps which should be taken to counteract them if they are deleterious to the race. The power of adaptation of the human being is very wonderful and it has operated through all history from prehistoric migrations to the opening up of the modern world and beyond to the present day. But the redistribution of the population of Europe over the tropical and sub-tropical parts of the globe has not been related to any scientific enquiry.[58]

Walter's approach, however, was not one shared with the leading members of the KSSRI, and it is in the attack on settler anxieties about the African climate and environment made by Kenyan eugenicists that the settler-biased political motivations of the movement are revealed.

Gordon rejected pessimism about the suitability of the Kenyan climate and the future of white settlement in Africa. What is more, by formulating his argument in eugenic terms, he linked the question of climate with white supremacy and political dominance. The *EAS* reported much of Gordon's lecture on 'What is Eugenics', given in 1933 to a 'particularly large audience' at the Theatre Royal in Nairobi, in which he broached the subject of climate and settlement:

> '... I can answer on greater authority than my own, that there is nothing known to science to prevent us from adapting ourselves if we desire it, to the climate of Kenya, as the best of our race adapted themselves to other climates, for example to those of England, Scotland, and Ireland! but that there is abundant knowledge in psychology to explain the fact that the so-

called nerves of Kenya are never due to climate but, as anywhere else, to the conditions in the individual solely the individual affair [sic].

'Even in more tropical countries than this, the grumblers about climate have always been those who had no intention of remaining in it. The same arguments were used about elevation and climate in the Transvaal until the success of the goldfields was established!'

With full confidence in the future of Kenya as a British Colony, and the coming triumph of those Europeans who make it their home, Dr. Gordon then invited his hearers to consider the education of their children with two axioms in mind: first that education was control of nurture to secure the best possible development of nature; second, the future of East Africa depended upon how ideals were maintained and imparted to the children.[59]

Here, Gordon explicitly linked a dismissal of the concept of the unsuitability of the African environment with a defence of the future of white settlement. He did this by referring to settler children, a potent symbol of the security and future of the European presence in the colony. Gordon's interest in the management of children was longstanding; in about 1909 he had a book published entitled *Modern Motherhood: A Guide to Girlhood, Motherhood and Infancy*, in which he gave detailed advice on the upbringing and welfare of girls.[60] Gordon's interest was clearly also widely shared by the KSSRI: the second 'immediate need' that the KSSRI saw for Kenya was the implementation of a 'Children Act for the protection of infant life and the prevention of cruelty to the young'.[61]

Dr Sequeira was particularly interested in European settler children. He was also influential in undermining fears among settlers about the medical effects of the African sun on Europeans. In an article in the *EAMJ* in 1932, Sequeira dismissed the commonly held theory of the dangers of actinic rays on European skin and argued that problems were more likely to be caused by overheating and dehydration, which were exacerbated by the clothing worn by Europeans.[62] He also argued that European children in Kenya were suffering as a result of anxiety about the effect of tropical sun on Europeans and that they actually needed more exposure to the sun.[63] Sequeira often spoke on European child health and development at the KSSRI, and he also emphasised the suitability of Kenya, in particular the White Highlands, for European settlement and for European children. He was more concerned with maternal responsibilities than climatic damage. The *EAS* reported him thus:

The appeal of this proposed Society was to people of character to build up character in the race. 'It is the function of the mother' said Dr Sequeira in an earnest peroration, 'to take charge of that most precious of all instruments, the product of countless ages of evolution, the infant brain. It is not a toy to be treated carelessly, to be neglected. It has in it potentialities for good or ill undreamt of ...'[64]

At a meeting of the Nakuru branch of the KSSRI, Sequeira again spoke on this matter, saying that the greatest concern arising from settlement in Africa was that European parents left too much of the care of their offspring to African servants, who lowered standards of intelligence and morality.[65] Sequeira was also vocal on the 'poor white' issue, a problem that he linked to his warnings about settler methods of parenting which relied heavily on the care given by African servants. Sequeira advocated bigger European families and the provision of scholarships to enable settler children to attend universities in Britain and South Africa.[66] Sequeira also used his editorials in the *East African Medical Journal* to warn of the dangers of degeneration of the white race, not because of climate, but through excessive contact with 'individuals of inferior civilization and culture', especially African house servants in charge of European children.[67]

The rejection of the idea that Kenya was an unsuitable place for European settlement was often turned into an attack on the relative degeneracy of the metropole:

> They are a healthy happy people & the children joyous sturdy little creatures living in a perfect climate with all the feed they require and beautiful surroundings
>
> I have not yet seen a deformity with the exception of a boy whose leg was lost in an accident
>
> Can we say the same in BRITAIN? Alas no or in India or any other part of the Empire.[68]

There was a political subtext to the eugenicists' defence of the suitability of Kenya's climatic conditions for Europeans. The biological soundness of white settlement became associated with the firmness of the European hold over the political situation in Kenya. The notion that Kenya might not be physically appropriate for white settlement was attacked by the eugenicists at the same time that settler political control of the colony was finally vetoed in the metropole.

Settler political aspirations received a severe blow in the form of the *Report of the Commission on Closer Union* of 1929, otherwise known as the Hilton Young Report after the Chairman of the Commission, the Conservative MP, Sir Edward Hilton Young.[69] Closer Union was supported by both Grigg and Amery, but the Cabinet insisted on a Royal Commission to consider the matter. As well as Hilton Young, the Commission contained Sir Reginald Mant, who had been a civil servant in India, Oldham and Sir George Schuster. The Commission divided between Hilton Young and the others, resulting in the production of minority and majority reports. The majority report dismissed the possibility of responsible government for the settlers and denied the European settlers an elected majority in the Legislative Council. It also reopened the question of franchise by referring to the civilisation test of enfranchisement and recommending a common roll of equal enfranchisement without racial discrimination.[70] Native paramountcy was re-emphasised. Hilton Young's minority report disagreed on the questions of responsible government and the common roll, but accepted native paramountcy. This report shocked and dismayed the settlers and in a telegram to Amery they threatened 'more vigorous action'.[71]

When the Closer Union report was published Gordon produced a long and detailed article in the *EAS*, in which he attacked the report and defended European supremacy. He began by questioning the notion that Kenya was not suitable as a 'white man's country'.[72] Gordon also questioned the report's emphasis on the capacity of East Africans to develop through 'education and civilisation', arguing that science indicated that mere environmental factors could not bring about such progress. Gordon also feared that greater equality might lead to miscegenation, which he saw as dangerous from a eugenic point of view:

> Known facts have allowed this confident generalisation: – That the crossing of stock belonging to the same main race (e.g. Irish and Scots, or English and German) produces a blended sub-race superior to the mean of the two parental stocks, or at least not inferior; but that the crossing of widely different stocks belonging to any two of the main four races (e.g. white and negroid) produces an inferior race.
>
> This I imagine clarifies the issue. As but one of the inevitable effects of the policy recommended by the Closer Union Commission, we are to become active agents in founding an inferior race.[73]

The responses to Gordon's article in the *EAS* correspondence columns are interesting. Two letters were published that disagreed with Gordon's approach to the Closer Union Report, one was written by 'Homo', a name used by a writer who was one of the most consistent critics of Gordon's theories in the letter pages of the Kenyan press.[74] The other was written by Sir Charles Foulis, who argued that Gordon's points were a side-issue given the fact that the European settler population was in the minority in East Africa and hence to change the political situation in their favour would be autocratic and oligarchic.[75]

A letter was also published in defence of Gordon, which gives a sense of the interest created by his article and the role of his theories in settler political thinking. 'B. S. C.' wrote:

> The [Gordon's] article is generally believed to have gone to the root of the matter. It examines impartially the assumptions on which the recommendations for native policy are based, points out the grave implications to our race in moderate language, and asks that before action is taken the truth of

these assumptions should be competently investigated ... For some years Dr Gordon has pleaded in a strictly impartial spirit for this investigation, and those who misrepresent his arguments and oppose his suggestions on no better ground than their own 'ideas', are open to the accusation that they wish to keep the truth unknown. His suggestions were publicly supported when originally made and were, I believe, brought before the Civil Research Committee only to be shelved pending the issue of the Hilton-Young Report.[76]

In this skirmish over the Closer Union Report, the political implications of Gordon's beliefs are expressed. When it came to the question of the effects of the Kenyan climate on Europeans and hostility towards racial equality, the political agenda of the Kenyan eugenicists was undisguised.[77] The underlying tension in the debate on climate was the permanence and integrity of European settlement in Kenya, which became linked to the question of whether Kenya could become a 'white man's country' politically. Gordon elaborated his view on the Kenyan environment in a letter to Nellie Grant about the biography of Lord Delamere, written by Grant's daughter, Elspeth Huxley, in which Huxley mentioned concerns about the effect of the climate on European 'nerves'.[78] Gordon wrote that:

> On the general question there can really be no doubt – that white man can & will adapt himself as he has done elsewhere <u>but</u> there will be rigorous biological selection from which – as elsewhere – as in America, Australia & so on – a modified white race will emerge.[79]

Gordon's interest in the future of white settlement in the face of racial competition seems to have been an important aspect of his public lectures on eugenics; it is likely that to many of the European settlers who attended Gordon's lectures, his pleas for strengthening the eugenic basis of Kenya's white population constituted the most important message. In his 1933 pamphlet 'Eugenics and the Truth about Ourselves in Kenya', which was based on a popular public lecture, Gordon declared:

> There is no question of more importance to the present and future of Kenya than the fall in the
> BIRTH RATE
> of *our own race,* and the liberal reproduction going on amongst other races here ... It is not necessary to urge here the importance of good fertility amongst those on whom the future of Kenya depends.[80]

By the time Lord Delamere died at the end of 1931 there was no longer any possibility of achieving an independent, settler-governed state and settlers had to concentrate on 'secondary objectives – financial control and ministerial office'.[81] The question of income tax thence became the major issue in settler politics until 1936. The effect of the Great

Depression on the Kenyan economy also meant that settlers were feeling particularly vulnerable in the early 1930s: by 1936 half of the general settler estates were not self-supporting and the total debt of the settler estate sector was estimated to be £3 million.[82] During the first half of the 1930s, then, the time when the Kenyan eugenics movement was at its peak, the Kenyan settlers were not only facing economic retrenchment, but also the fact that their longstanding demand for political control was not going to be met. In other words, it was a period in which settler ambitions in Kenya had to be renegotiated around economic and political constraints that undermined the feasibility of their claims for Kenya. The formation of the Kenyan eugenics movement, about twenty to thirty years after its inception in Britain, occurred, therefore, at a time when the settler population was receptive to eugenic consolation and biological solutions to a political predicament.

African education

The controversy around the question of race and intelligence and the educable capacity of the 'East African native' coincided with the emergence of the Kikuyu Independent Schools Movement. Most of the schools for Africans were run by missions: the average enrolment of Africans at mission schools was 82,455. Some of the mission schools were assisted by government funding; the overall expenditure on African and Arab education in 1929 was £74,043. In the same year, just under 1,500 Europeans attended government and non-government schools, with an expenditure of £49,993.[83] By 1935, there were on average 3,993 Africans enrolled at government schools and 92,586 at mission schools, costing £74,097 (expenditure on Arab education was by this time counted separately). Average European enrolment at government schools was 1,121 and 687 at non-government schools, with an expenditure of £44,041.[84] This considerable inequality in the provision of education was being increasingly disputed by the African population in Kenya in the 1930s; the KSSRI can be seen as confronting this by making African educability central to the colonial eugenic problematic.

Overarching metropolitan policy on African education had been set forth in a 1925 memorandum produced by the Advisory Committee on Native Education in the British Tropical African Dependencies.[85] The memorandum argued that official involvement and encouragement should be increased and African education should no longer be left entirely in the hands of the missions. The aims of education were to include:

> the raising up of capable, trustworthy, public-spirited leaders of the people, belonging to their own race. Education thus defined will narrow the hiatus between the educated class and the rest of the community, whether chiefs or peasantry ... The first task of education is to raise the standard alike of character and efficiency of the bulk of the people, but provision must also be made for the training of those who are required to fill posts in the administrative and technical services, as well as those who as chiefs will occupy posts of exceptional trust and responsibility. As resources permit, the door of advancement, through higher education, in Africa, must be increasingly opened for those who by character, ability and temperament show themselves fitted to profit by such education.[86]

The emphasis placed on the role of education in African development and the underlying assumption of both the validity and necessity of such social progression were clearly out of tune with the nature of African educability conceived by the Kenyan eugenicists. What is more, the general principles set forth in the 1925 memorandum were accepted and reiterated in the report on Closer Union of 1929.[87] The issue of African education was intimately linked with social and political advancement; this politicisation of education is crucial to understanding the contemporary political meaning of the Kenyan eugenic theories. The development of the independent schools movement further heightened the charge of the politics of education in Kenya.

Tension had arisen over the control of education by missions in the out-schools disputes of the 1920s, with an independent school being founded as early as 1922 in the village of Gituamba in Fort Hall.[88] The clitoridectomy crisis of 1929 hastened the movement for the separation of education from missions in central Kenya.[89] The Church of Scotland Mission (CSM) opposed clitoridectomy with the most vehemence, particularly J. W. Arthur, head of the mission at Kikuyu. The CSM insisted that its adherents oppose female circumcision; this was refused by 90 per cent of the church members. This conflict seriously affected the educational work of the mission: out-school attendance fell from 'an average of 728 in December 1928 to only eighty-seven a year later', with parents refusing 'to accept as teachers those who had declared their loyalty to the CSM'. The Africa Inland Mission, which also took a firm line on female circumcision, had twenty of its out-schools closed in the Fort Hall district by 1930.[90] Most Church Missionary Society (CMS) missions were more tolerant of the practice; the one CMS station that took a hard line, John Comely at Kigari in Embu, suffered almost immediate consequences.[91]

The Independent Schools Movement really took off in the early 1930s. The Kikuyu Independent Schools Association (KISA) was formed, which established its own schools that eventually accepted government inspection in return for grants in aid, and the movement was also closely associated with the formation of independent churches. KISA was less overtly political than another independent schools organisation, the Kikuyu Karinga Education Authority (KKEA), which was confined to southern Kiambu.[92] KISA had schools as far afield as the Rift Valley Province and northern Tanganyika, although it was concentrated in Fort Hall, Nyeri and northern Kiambu.[93] It is striking that the Kenyan eugenicists started developing their theories on African educable capacity at a time when education was an important element of the Kikuyu political agenda and became associated with a political movement that profoundly threatened the colonial status quo. The independent schools movement connected access to education with access to social, political and economic objectives, indicated in particular by the KKEA's clear anti-colonial stance.

The issue of the nature of the provision of education for Africans was always difficult because of unease about the links between education and political subversion. There was a distrust of what was termed 'literary' education and the teaching of English; the language in which African schooling was conducted became politicised as proficiency in English was seen as a further breach of the racial barrier. The KISA schools taught in English at all standards,[94] whereas in other mission and government schools English was, rather grudgingly, only taught at the highest stages. In 1936 KISA attended a meeting with District Commissioners and the Provincial Commissioners at the Jeanes School, where it was agreed that English instruction would begin at Stage Three and that English would become the medium of instruction at Stage Four.[95] African political unrest was often blamed on inappropriate education, with comparisons frequently made to British schoolboys who got carried away with ideas.[96] The appearance of independent schools strengthened the perceived connection between education and politics; at a meeting of Provincial Commissioners in reference to the female circumcision crisis it was recommended that 'propaganda to counteract agitation subversive to Government should be undertaken more especially in the schools among the young natives of the Colony'.[97] The settler perception of the dangers arising from the political situation in 1930, and the role of education in explaining it are captured in the following letter to the Colonial Office:

> We are living in troublesome times, and a Kikuyu rising is imminent. Settlers are beginning to hold meetings, and a Vigilance Committee has been formed. A settler from Nyero has been up, and sounded an urgent note of warning, and says that we have no idea how bad it is. He has warned us that if any of us have any young children, that we are not to let them out of our sight for a moment. Women all over Kiambu have been warned by friendly natives never to be alone for a moment. We are supposed to have

the visiting Bolshy Organizers on our, and surrounding farms. For the last six months all the young Kikuyus have been uppish, and cheeky, and have refused to turn out for work. *It appears that the more educated they are, the more they seem to be in it* ... There has been another episode recently near Nairobi concerning two white women being dragged off their horses. The position out here is atrocious. Things are really very black in this country, and frankly I do not like our position here at all, as we are surrounded by so many Kikuyus.[98]

Settler concern about Kikuyu political activity was clearly intense in 1930 and education was considered a cause of subversion; it was associated with youths who rejected the traditional, rural status quo.

One of the founders of KISA, Parmenas G. Mockerie, attempted with Kenyatta to give evidence as a delegate of the KCA before the Parliamentary Committee on East Africa in Britain in May 1931 and then went on to study at Fircroft College, Birmingham, and Ruskin College, Oxford. On his return to Kenya, Mockerie was found to possess subversive literature which was confiscated.[99] At this time, Kenyatta was also studying at the LSE. This context of the well-known educational aspirations and achievements associated with Kikuyu political leaders who aroused so much hostility and fear among European settlers must be borne in mind when considering the assertions of Kenyan eugenicists about the educable capacity of East Africans. In his book, *An African Speaks for His People*, Mockerie referred to the issue of mental capacity in terms of the political issues surrounding African education and political rights:

> The reason given for refusing to Africans direct representation in the Government is that the Africans have not sufficient education to enable them to take an intelligent part in councils composed of persons who have behind them a culture of centuries, to whom the opportunity of higher education has been afforded. It may be suggested that education which will suit Indians and Europeans is not applicable to the mental capacity of Africans. This is a suggestion which will not be accepted by Africans, who are conscious of the necessity of having a sound education similar to that given to children of other races. Why should a Government which has responsibility for educating the children of the people whom it taxes give to some better opportunities than to others? It is unjust that when the official language is English, opportunities should not be given, but rather difficulties put in the way of Africans who wish to learn English ... It ... makes the ruling class feel that the Government is its own, while the subjected race is compelled to think of the Government as alien.[100]

The coincidence of education becoming one of the most controversial Kikuyu political complaints and the elaboration and success of the eugenic theories about African intelligence from about 1930 to 1937 is striking. The following letter from a settler in Nakuru conveys this racial anxiety and paranoia, bringing in the common assertions of the need for European political intransigence and the dangers of not taking into account racial inferiority in intelligence:

> defence association have protested most strongly to the Governor that it is high time that this HUSH HUSH was put a stop to about these appalling crimes that the danger of the White Colony were known in England where it is the fashion to abuse the White man & pamper the Black
>
> If Bloodshed occurs it will be entirely the fault of the inane policy of attempting to rule a savage black race from below instead of on the top …
>
> The trouble in India at the present must surely make those at home in charge of affairs realize that they cant [sic] play with mentally inferior people.[101]

Gordon legitimised and rationalised the settler suspicion of African education. He often described excessive faith in education in colonial development as a curse of modernity: 'In one way the Romans were more fortunate than ourselves (in their Colonial Empire); they had no glib adviser of moderation, none to slander them if they were not moderate, none possessed by blind faith in the omnipotence of the blessed word, Education …'[102]

As Dr Sequeira argued in reference to the theory on African mental capacity, the provision of conventional education was potentially disastrous: 'It has long been recognised that educational methods applied to normal children cannot be applied to the backward and defective.'[103] The presentation of the ideas of the Kenyan eugenicists, then, occurred at a time when the issue of education was particularly politically heated, and the question was directly related to the security of the position of Europeans. Education was linked to white prestige, as well as African political subversion; indeed African education was potentially subversive precisely because it encroached on white prestige. In 1933 a cartoon was published in the *East African Weekly Times* that showed a caricature of two foppish and insolent young black men in a quintessential Oxbridge setting, one wearing a monocle and one smoking a cigar, mocking the generosity of 'Trusteeship'. The caption beneath read: 'European education in Kenya threatened by the proposal to increase tuition fees for secondary schools and by the absence of compulsory primary education.'[104]

In 1934 education was made compulsory for Europeans after an ongoing debate about the development of a class of poor whites, an issue that Francis Scott, in the Legislative Council, directly related to the permanence and authority of the European community:

> if it were accepted that white settlement was here for good it was essential

the white settlers should be the best equipped people. The people who would be the dominating influence of the future were the children of to-day, and it was absolutely essential for them to have the best education. It was important, too, to the native population that the white children be properly educated, so that there would be no fear of an illiterate class.[105]

The Advisory Council on European Education put the issue in terms of the labour market and the fact that the large supply of cheap African labour made the emergence of unskilled European labour impracticable. Hence each European should take the highest possible position in employment that was 'commensurate with his capacity' and an adequate education should be provided that would enable the European to reach the highest possible intellectual capacity.[106] It is significant that compulsory European education was implemented at a time when African education was so politicised, and when the eugenicists were also placing education at the forefront of their agenda. The eugenicists made education and educability a defining factor in racial difference. This heightened the need to enforce European educational standards at a time when economic recession had caused settler juveniles to be withdrawn from schools.[107]

In his memoirs, Trowell made the link between Gordon and Vint's theories and the settler political agenda:

> Undoubtedly a certain group in Kenya wanted to believe that the future of the country lay in White Settlement in the Highlands: Africans would remain hewers of wood and carriers of water to the end of time: another South Africa. Other settler and many commercial Europeans took a more liberal view and encouraged the education and development of Africans, but they were a small minority.[108]

The extent of support for Gordon's attitude to African mental capacity was indicated by the response given to Gordon's dismissal as Visiting Physician to Mathari Mental Hospital. Not only did the Kenya BMA pass a resolution urging that Gordon be retained, but Major Cavendish-Bentinck, one of the most extreme settler leaders, also became involved, demanding that Gordon at least be paid a pension, despite the fact that his post as Visiting Physician did not warrant one.[109] The Legislative Council instead awarded Gordon with an honorarium of £100 for his 'valuable service'.[110] In the late 1930s, despite the demise of the KSSRI, the failure of the campaign for the establishment of the research programme into race and intelligence, and his dismissal from Mathari, Gordon continued to give public lectures and attract large audiences,[111] warning of the dire effects of mental degeneracy throughout the colony.[112]

Officials and the eugenic research

As shown in Chapter 4, officials in Nairobi were in regular contact with the Colonial Office in London on the matter of race and intelligence and at times helped to restore flagging metropolitan government interest in the campaign. Here the official response in Nairobi will be examined as part of the story of European involvement with eugenics within the colony. While colonial officials in London regarded the Kenyan research with wary detachment that developed into outright scepticism, those in Nairobi were in a more difficult position. Officials in Kenya generally had more misgivings than the unofficial European settlers, but were less likely to disparage scientific racism than their colleagues in the metropole. The European population in Kenya was small and close knit (16,812 according to the 1931 census); despite some settler hostility towards officials, inevitably the two groups were closely connected socially and culturally. Official responses to the Kenyan eugenicists were either moderated by the need to maintain good relations with settlers or because the officials concerned were genuinely interested in eugenics as part of a progressive approach to African development and welfare. The individuals who showed an active interest in eugenics will be discussed first. This will be followed by an examination of the responses of the highest levels of the administration in Kenya. Given the considerable interest aroused by the brain research, successive governors and colonial secretaries could not avoid becoming involved in the issue; the nature of the support these officials conferred will be analysed.

Despite the appeal of eugenics to certain reactionary settler political interests, official support was often motivated by more liberal, social hygienist intentions. There were members of the KSSRI who were identified by contemporaries in the colony as being social progressives. Key proponents of this were La Fontaine and Paterson. The more 'pro-native' individuals involved in the KSSRI were mostly from the administrative rather than the non-official settler section of the European community. In particular, they were members of the colonial administration who tended to be involved in 'native affairs' and likely to consider their responsibilities to include the paternalistic protection of native interests in the face of settler excesses.

Individual departments within the Kenyan administration acknowledged the research of Gordon and Vint. Gilks and Paterson, Directors of Medical Services, and Scott, Director of Education, were all involved in the campaign for research into 'native backwardness'. Gilks had retired by the time the campaign for research really got underway and was usefully placed in Britain as an agitator on behalf of the Kenyan eugenicists. La Fontaine, who was at that time Acting Chief Native Commissioner, was also a supporter and a high-level official presence at KSSRI meetings. Paterson and Scott both gave their support to the campaign in their official capacities through their annual reports. Given the publicity surrounding the issue of race and intelligence, it was necessary for them to make statements on the question. This was recognised in the Colonial Office in London when there was some disagreement between Paterson and the Governor over the length and content of Paterson's Annual Medical Report for 1934.[113]

In his report, Paterson discussed the question of mental capacity at length, beginning with the statement: 'It is common experience that many Africans are, at least by European standards, dull'.[114] Paterson then went on to state the importance of eliminating disease and other environmental barriers to African development. He then mentioned the brain research, emphasising its preliminary nature and the need for more investigation into African mentality and physiology. Flood wrote that Paterson's comments were:

> really required in present circumstances, in view of the hullabaloo that is going on as a result of Dr Gordon's efforts. It would be impossible for Dr Paterson to pass over those efforts in his Report, and what he says represents his own attitude towards the problem that has been raised by Dr Gordon.[115]

As Director of Education, Scott also responded to the matter in his Annual Report:

> The subject is admittedly controversial, but it may be confidently asserted that the interest aroused by Dr. Vint's paper was widespread and deep. It is evident that further investigations in the field of research would be of great value to those engaged in the education of the native, and it is earnestly to be hoped that the line of inquiry which has been started will not be abandoned through lack of the necessary funds.[116]

Scott, as shown in Chapter 3, was one of the few voices in Kenya who presented the environmentalist case, questioning the extremism of some of the conclusions presented by Gordon and Vint. Yet Scott was still a supporter of the need for the foundation of a more extensive research project on mental backwardness. In fact, as far as he was concerned, the significance of environmental factors meant that there was a need for an even larger research project that could encompass an investigation of the 'social conditions of the African'. He wrote to the Colonial Secretary:

> I am of the opinion that this inquiry is likely to be of great value if it is carried out on a sufficiently wide front.
> 2. It seems to me essential that the histological investigations of Dr Vint should be associated with the psychiatrical [sic] psychological investiga-

tions, together with definite enquiry into the social conditions of the African.

3. The fears ... that the result of such an investigation might be politically embarrassing do not appear to me to be a ground for burking the investigation. If the inferiority of the average African is established then the problem of selection in regard to educational and social advancement becomes of vital importance to the Educationist. My arguments as against the conclusions of Dr Gordon are that they are based on averages and we have to find out to what extent there is an overlap. If for every 100 Europeans who are fit to be sent to Cambridge or Oxford there are 5 natives, then we should still need our Makerere [the university in Uganda]. Booker Washington and Agrey are facts.[117]

At the highest levels of the Kenyan administration, there was also support for the work on 'native backwardness'. Governor Sir Edward Grigg, the Colonial Secretary Sir Armigel Wade (whilst Acting Governor), and Governor Sir Joseph Byrne all praised Dr Gordon in the Kenya Legislative Council, although the intensity of their support varied.[118] The real problem was the economic retrenchment in Kenya in the 1930s, which precluded any government expenditure on a major research programme and meant that official support in Kenya was powerless without metropolitan backing. The economic crisis had put other pressures on the government. Under the Agricultural Ordinance of 1930, £113,000 was lent to farmers by 1934; £143,000 was spent on refunds and rebates between 1930 and 1933; and loan subsidies of £116,000 were granted on exports of the 1929–39 cereal crop.[119] The government had to cut back on expenditure on administration: in these circumstances the provision of funding from within the colony for the research into African intelligence was not feasible.

Edward Grigg was particularly influential in encouraging Gordon to embark on the research into 'native backwardness'. Gordon himself wrote that Grigg first suggested the need for scientific investigation in this area in 1926:

> The direct suggestion that help from science should be obtained in East Africa came in 1926 from a member of this circle [the African circle at Chatham House, London], Sir Edward Grigg, whose words reached me under a sausage tree at Koru, where I was speculating about the causes of backwardness ... Sir Edward's words turned me into a public agitator ...[120]

The idea of scientific colonisation appealed to Grigg, who was 'imbued with a grandiose imperialist vision'.[121] Grigg and Gordon maintained friendly relations after Grigg left Kenya; in 1934 he arranged to have the printing of Gordon's population *Atlas* paid for.[122]

Grigg retired as Governor in 1930 in a state of extreme irritation with the Labour government over the questions of settler self-government, native paramountcy and closer union. The loss of Grigg and his enthusiastic support was a blow to Kenyan eugenics. Having started his career a Liberal, Grigg became a staunch Conservative during his tenure as Governor (1925–30) and became increasingly identified with settler leaders such as Lord Delamere and sympathetic to their political aspirations and their attitudes on race. Grigg wrote in 1931: 'I believe that race is an essential element in the government of the world, and that nothing but harm can come from failure to back our own race. That, you will remember, was part of Lord Milner's written creed. It has always been part of mine.'[123] Grigg was a good friend of Amery, Secretary of State for the Colonies, 1924–29, who was a proponent of eugenics and shared Grigg's tendency to favour settler over African interests. Grigg, Delamere and Amery shared similar visions of a settler-dominated Closer Union in East Africa. Although Byrne, who succeeded Grigg, and Byrne's Colonial Secretaries, Moore (whose wife was on the committee of the KSSRI) and Wade, all expressed their support for the research on race and intelligence, the favourable official environment that developed under Grigg was not recreated to the same extent in the 1930s.

The death of Lord Delamere in 1931 and the demise of the idea of Closer Union and settler ambitions for responsible government all marked a change in emphasis in the Kenyan political situation. Joseph Byrne, the new Governor, represented a new regime that was less complicit with settler attitudes and aspirations. Sir Francis Scott, who replaced Delamere as the main leader of the settlers was more moderate (although his role was challenged by the more extremist Cavendish-Bentinck); Scott used his powerful social and political connections in the metropole to attempt to improve relations with the Colonial Office. As Bennett put it: 'With this new leadership in the Colonial Office and Nairobi, the temper of politics changed – at least at the summit.'[124]

Given the strength of the consensus within the Kenyan medical profession concerning African backwardness and the resonance of these views with those of settlers more generally, it would have been difficult for any governor not to have given some support to the eugenic campaign. It should also be borne in mind that at this time eugenic thinking was widespread within the British ruling classes; that there was a consonance between the ideas about race and eugenics expressed by Gordon and his colleagues in Kenya and the ideas about race and class that were likely to be held by most members of the official class. The levels of hostility and agitation that settlers were quick to mobilise against the administration meant that officials, particularly a senior and longstanding figure like Wade, found service in Kenya demanding (Sir Armigel Wade had been Chief Native Commissioner, taking up the post of Colonial Secretary in

1934).[125] Vouching some sort of support for the research on mental capacity was probably considered an unusually easy opportunity to meet a settler demand; it was in keeping with settler opinion, as expressed in the *EAS*, and it was relatively uncontroversial.

The British government argued that the support for the brain researches within Kenya was unofficial and, because of its political implications, preferred it to remain so.[126] Joseph Byrne generally conformed to this stance. It appears that Byrne was walking a tightrope between metropolitan official scepticism and settler enthusiasm. He did express admiration for the research in the Legislative Council[127] and in 1935 he accepted an invitation to hear Gordon read his paper on 'An Inquiry into the Correlation between Civilization and Mental Disorder in the Kenya Native' at a meeting of the Kenyan branch of the BMA.[128] In July 1934 Byrne wrote a despatch stating that he had 'no doubt that the institution of research on these lines would help to solve many of the administrative, educational, social and political problems'.[129] Byrne also became involved in the campaign to keep Vint when it was proposed that he move to Mauritius; he sent a long telegram and a dispatch to the Colonial Office strongly urging that Vint be retained in Kenya.[130] Yet while Grigg truly embraced the settler stance on eugenics and race, there is no evidence of a similar level of enthusiasm emanating from Byrne, suggesting that he was motivated by diplomatic expediency rather than genuine commitment to the eugenic programme.

Overall, official responses within Kenya in the 1930s to the theories on 'native backwardness' were either discreetly positive, as in the case of Byrne, or motivated by a reformist agenda, as with Paterson. Officials tended to emphasise the preliminary nature of the research and the need to take into account the influence of environmental factors more than settler eugenicists did, reflecting the fact that officials were more likely to fall into the meliorist, social hygiene wing of Kenyan eugenics. Despite this difference in approach, the fundamental validity and utility of the research were not questioned and this more temperate support from high-level government officials reinforced the image of reasoned objectivity presented by the Kenyan eugenicists. The real obstruction to the development of more research lay in the external factors of economic depression and dependence on the metropole which meant that the Kenya administration could not provide the financial support that was required.

Notes

1 'To Study Improvement of the Race', *EAS*, 3 June 1933, p. 11.
2 'The Improvement of the Human Race', *EAS*, 25 March 1933, pp. 14–15.

3 'To Study Improvement of the Race', p. 11.
4 Gordon to Moore, 25 March 1933, CMAC, SA/EUG/C.129.
5 'To Study Improvement of the Race', p. 11.
6 Ibid.
7 Grant to Huxley, 12 July 1933, RH, MSS.Afr.s.2154/1/1.
8 G. R. Searle, 'Eugenics and Politics in Britain in the 1930s', *Annals of Science*, 36 (1979), p. 160.
9 D. Kennedy, *Islands of White: Settler Society and Culture in Kenya and Southern Rhodesia, 1890–1939* (Durham, 1987), p. 197.
10 Grant to Huxley, 12 July 1933, RH, MSS.Afr.s.2154/1/1
11 Searle, 'Eugenics and Politics in Britain in the 1930s', p. 160.
12 Grant to Huxley, 12 July 1933, RH, MSS.Afr.s.2154/1/1
13 Grant to Huxley, 9 November 1933, RH, MSS.Afr.s.2154/1/1
14 Grant to Huxley, 28 September 1933, RH, MSS.Afr.s.2154/1/1
15 E. Huxley, *White Man's Country: Lord Delamere and the Making of Kenya*, Volumes I and II (London, 1935). Huxley debated the settler position in her correspondence with Margery Perham, which was then published in E. Huxley and M. Perham, *Race and Politics in Kenya* (London, 1944).
16 E. Huxley (ed.), *Nellie. Letters from Africa* (London, 1980).
17 Gordon to Mrs Collyer, 26 September 1938, CMAC, SA/EUG/D.69.
18 'To Study Improvement of The Race', p. 11.
19 Grant to Huxley, 14 December 1933, RH, MSS.Afr.s.2154/1/1.
20 KNA, Health/1/2980.
21 Grant to Huxley, 12 July 1933, RH, MSS Afr.s.2154/1/1.
22 'Report on the Applicability to Kenya of Methods Pursued in Borstal and Other Reformatory Schools in England', NA, CO 533/426/17 (also in KNA, AP/1/701).
23 KNA, Secretariat/1/7/9.
24 T. Askwith, *Memoirs of Kenya, 1936–1961*, p. 25, RH, MSS.Afr.s.1770 (1).
25 T. Askwith, *From Mau Mau to Harambee: Memoirs and Memoranda of Colonial Kenya* (Cambridge, 1995), p. 48.
26 Grant to Mrs Collyer, 26 September 1938, CMAC, SA/EUG/D.69.
27 Kenya Society for Race Improvement invitation card, 15 May 1933, CMAC, SA/EUG/C.129.
28 'Kenya Society for the Study of Race Improvement', *EAMJ*, 10/4 (1933), p. 155.
29 Grant to Huxley, 24 October 1934, RH, MSS.Afr.S.2154/1/2.
30 *The Mission Hospital*, 1930–36.
31 Gordon to Blacker, 5 October 1930, CMAC, SA/EUG/C.129.
32 'Inquirer' to *EAS*, 8 July 1933, p. 35.
33 'George Willie' to the *EAS*, 28 May 1932, p. 36; see also H. S. Scott, 'A Note on the Educable Capacity of the African', *EAMJ*, 1 (1932), p. 581.
34 Grant to Huxley, 16 January 1935, RH, MSS.Afr.s.2154/1/1.
35 'To Study Improvement of the Race', p. 11.
36 M. G. Redley, 'The Politics of a Predicament: The White Community in Kenya 1918–1932', PhD thesis (Cambridge, 1976), p. 14.
37 'The Study of Race Improvement', p. 46.
38 C. J. D. Duder and C. P. Youé, 'Paice's Place: Race and Politics in Nanyuki District, Kenya, in the 1930s', *African Affairs*, 93 (1994), pp. 253–78, 266.
39 'The Study of Race Improvement', p. 46.
40 Ibid.
41 A. Davis to *EAS*, 20 January 1934, p. 36.
42 R. F. Mayer to *EAS*, 27 January 1934, p. 36.
43 Kisosonkole to *EAS*, 21 May 1932, p. 36.
44 Kisosonkole to *EAS*, 6 October 1934, p. 35.
45 Kennedy, *Islands of White*.
46 Grant to Huxley, 17 October 1934, RH, MSS.Afr.s.2154/1/2.
47 H. Hinde, 'The "Black Peril" in British East Africa. A Frank Talk to Female Settlers', *The Empire Review and the Journal of British Trade*, 35/245 (1921), pp. 193–200.

48 S. Klausen, '"For the Sake of Race": Eugenic Discourses of Feeblemindedness and Motherhood in the South African Medical Record, 1903–1926', *Journal of Southern African Studies*, 23/1 (1997), pp. 27–50, and S. Dubow, *Scientific Racism in Modern South Africa* (Cambridge, 1995).
49 Dubow, *Scientific Racism in Modern South Africa*, p. 184.
50 J. Hyslop, 'White Working-class Women and the Invention of Apartheid: "Purified" Afrikaner Nationalist Agitation for Legislation Against "Mixed Marriages", 1934–9', *Journal of African History*, 36/1 (1995), pp. 57–81.
51 Gordon to Grant, 6 October 1935, enclosed in a letter from Grant to Huxley, 23 October 1935, RH, MSS.Afr.s.2154/1/3.
52 Kennedy, *Islands of White*. See Chapter 6, 'A Climate of Concern', pp. 110–27.
53 Ibid., p. 115.
54 M. Mackinnon, 'Medical Aspects of White Settlement in Kenya', *EAMJ*, 11/12 (1935), pp. 376–93, 388. See also D.V. Latham, 'The White Man in East Africa', *EAMJ*, 9/10 (1933), pp. 276–82.
55 A. Walter, 'Climate and White Settlement in the East African Highlands', *EAMJ*, 11/ 7 (1934), pp. 210–25.
56 Ibid.
57 Editorial, 'Climate and White Settlement', *EAS*, 19 May 1934, p. 43, and Editorial, *EAMJ*, 11/ 7 (1934), pp. 209–10.
58 Editorial, 'Climate and White Settlement', p. 43.
59 'The Improvement of the Human Race', pp. 14–15.
60 H.L. Gordon, *The Modern Mother: A Guide to Girlhood, Motherhood and Infancy*, (London, ?1909).
61 'The Improvement of the Human Race', p. 14.
62 J. H. Sequeira, 'The influence of Light and Heat on the Human Body', *EAMJ*, 8/12 (1932), pp. 332–58.
63 Grant to Huxley, 28 September 1933, RH, MSS.Afr.s.2154/1/1.
64 'To Study Improvement of the Race', p. 11.
65 'The Rising Generation in Kenya', *EAS*, 18 November 1933, p. 33.
66 'The Rising Generation in Kenya', p. 33, and 'Study of Race Improvement', p. 46.
67 Editorial, *EAMJ*, 9/10 (1933), p. 275.
68 Letter from A. Loch, Nakuru, 6 May 1930, NA CO/533/398/11.
69 *Report of the Commission on Closer Union of the Dependencies in Eastern and Central Africa*, Cmd. 3234 (London, 1929).
70 *Report of the Commission on Closer Union*. See also G. Bennett, 'Settlers and Politics in Kenya up to 1945' in V. Harlow and E. M. Chilver (eds), *Oxford History of East Africa* (Oxford, 1965), pp. 309–10.
71 Cited in Bennett, 'Settlers and Politics in Kenya', p. 310
72 H. L. Gordon, 'Some Implications of the Closer Union Report', *EAS*, 13 April 1929, pp. 15–16.
73 Ibid.
74 'Homo' to *EAS*, 13 April 1929, p. 36.
75 Foulis to *EAS*, 15 April 1929, p. 33.
76 'B. S. C.' to *EAS*, 27 April 1929, p. 33.
77 See 'The Improvement of the Human Race', pp. 14–15.
78 Huxley, *White Man's Country*, vol. 2, p. 299.
79 Gordon to Grant, 6 October 1935, enclosed with a letter from Grant to Huxley, 23 October 1935, RH, MSS.Afr.s.2154/1/3.
80 Gordon, 'Eugenics and the Truth about Ourselves in Kenya', pp. 5–6, NA, CO 822/55/1.
81 Bennett, 'Settlers and Politics in Kenya'.
82 B. Berman, *Control and Crisis in Colonial Kenya: The Dialectic of Domination* (London, 1990), p. 166.
83 Education Department Annual Report, 1929, pp. 81–9.
84 Education Department Annual Report, 1935, pp. 71–8.

85 *Memorandum on Education Policy in British Tropical African Dependencies*, Cmd. 2374 (London, 1925).
86 Ibid., p. 4.
87 *Report of the Commission on Closer Union of the Dependencies*, p. 72.
88 J. B. Ndungo, 'Gituamba and Kikuyu Independency in Church and School' in B. McIntosh (ed.), *Ngano, Studies in Traditional and Modern African History* (Nairobi, 1969), pp. 131–50.
89 Ndungo, 'Gituamba and Kikuyu Independency in Church and School' and J. M. Lonsdale, 'The Moral Economy of Mau Mau' in B. Berman and J. M. Lonsdale (eds), *Unhappy Valley: Conflict in Kenya and Africa, Book Two: Violence and Ethnicity*, (Oxford, 1992), pp. 265–504. See also L. M. Thomas, 'Imperial Concerns and "Women's Affairs": State Efforts to Regulate Clitoridectomy and Eradicate Abortion in Meru, Kenya, c. 1910–1950', *Journal of African History*, 39/1 (1998), pp. 121–45.
90 R. W. Strayer and J. Murray, 'The CMS and Female Circumcision' in R.W. Strayer (ed.), *The Making of Mission Communities in East Africa: Anglicans and Africans in Colonial Kenya, 1875–1935* (London, 1978), pp. 136–55, 139–40.
91 Ibid., p. 147.
92 Lonsdale, 'The Moral Economy of Mau Mau', p. 395. For more on the Kikuyu Central Association, see J. Spencer, *The Kenya African Union* (London, 1985), Chapter 3, 'Background to KAU II: The KCA Years 1922–40', pp. 55–114.
93 A. S. Adebola, 'The London Connection: A Factor in the Survival of the Kikuyu Independent Schools Movement, 1929–39', *Journal of African Studies*, 10/1, pp. 14–23; Ndungo, 'Gituamba and Kikuyu Independency in Church and School'.
94 Rules of the Kikuyu Independent Schools Association, p. 2, KNA, Secretariat/1/7/9.
95 Ndungo, 'Gituamba and Kikuyu Independency in Church and School', p. 136.
96 For example, Letter from DC, Fort Hall, to PC, Nyeri, 29 January 1930, KNA, PC/CP/8/7/1.
97 Cited in a dispatch from Grigg to Passfield, 21 March 1930, NA, CO/533/398/11.
98 Letter, 11 February 1930, enclosed in correspondence from Somerville to Shiels, NA CO/533/398/11. My italics.
99 Confidential letter from Commissioner of Police to the Provincial Commissioner, Fort Hall, 3 December 1935, KNA, Secretariat/1/7/9.
100 P. Mockerie, *An African Speaks for His People* (London, 1934), p. 59.
101 Letter from A. Loch, Nakuru, 6 May 1930, NA CO/533/398/11.
102 'Value of a Healthy Group Spirit', *EAS*, 16 April 1932, p. 31.
103 J. H. Sequeira, 'The Brain of the East African Native', *EAS*, 30 April 1932, p. 45.
104 'Is This Why European Education is Neglected?', *East African Weekly Times*, 8 December 1933, p. 17.
105 'Legislative Council Debate on Education', *EAS*, 22 July 1930.
106 Advisory Committee on European Education, Memorandum on Education Policy, 27 February 1934, KNA, Education/1/962.
107 Education Department Annual Report, 1931, pp. 21–2.
108 Trowell, *Memoirs*, p. 9, RH, MSS.Afr.s.1872, Box XXXIV.
109 Note by Colonial Secretary, 28 June 1937, KNA, AG/32/231.
110 KNA, AG/24/28.
111 Gordon to Blacker, 16 July 1939, CMAC, SA/EUG/C.130.
112 'Mental Degeneracy', *EAMJ*, 16/4 (1939), pp. 190–1
113 NA, CO 533/440/7.
114 Medical Department Annual Report, 1933, p. 31.
115 Comment by Flood, 17 October 1934, NA, CO 533/440/7.
116 Education Department Annual Report, 1932, pp. 2–3, NA, CO/544/38.
117 Scott to the Colonial Secretary, 3 February 1934, KNA, BY/26/7.
118 'Dr Gordon's Dismissal', 11 February 1937, NA, CO 822/78/19.
119 C. C. Wrigley, 'Kenya: The Patterns of Economic Life, 1902–1945' in V. Harlow and E. M. Chilver, *Oxford History of East Africa* (Oxford, 1965), pp. 209–64, 248.
120 H. L. Gordon, 'The Mental Capacity of the African', *Journal of the African Society*, 33/132 (1934), p. 227

121 Kennedy, *Island of White*, p. 75.

122 Gordon to Grigg, 12 January 1934, Bodleian, Mss.Film.1003.

123 Grigg to the Hon. R. H. Brand, 6 August 1931, Bodleian, Mss.Film.1003.

124 Bennett, 'Settlers and Politics in Kenya', p. 318.

125 The frustration and stress Wade himself felt after service in Kenya is described in a letter cited by Berman, *Control and Crisis in Colonial Kenya*, p. 139.

126 Note by Ackeson, 17 September 1932, NA, CO 822/47/5.

127 'Dr Gordon's Dismissal', 11 February 1937, NA, CO 822/78/19.

128 Letter from Bird (Private Secretary to the Governor) to Trowell (Hon. Sec. of the Kenyan BMA), 13 November 1935, KNA, GH/7/30.

129 Dispatch from the Governor (Byrne) to Cunliffe-Lister, 5 July 1934, NA, CO 822/61/14.

130 Comment by Flood, 23 March 1937, NA, CO 822/78/19.

CHAPTER 6

Biology, development and welfare

In this chapter, the effects of biological thinking on attitudes towards African development and social policy in Kenya will be explored using juvenile delinquency, intelligence testing and mental health as examples. Debates about juvenile delinquency and criminal insanity were domestic aspects of a wider eugenic debate about African educability and social progress, but the colony also fed into an international circuit interested in race and intelligence through research conducted under the auspices of the Carnegie Corporation.

But first of all, any discussion of eugenics and social policy needs to take into account the limits of the Kenyan state. As Joanna Lewis has demonstrated, the Kenyan colonial state was peculiarly spartan in its provision of social welfare and community medicine.[1] The problem for anybody with a eugenic agenda in Kenya and wishing to implement eugenic policy measures along the lines of those aimed at by eugenicists in Britain was that the infrastructure by which the social problem group could be traced did not exist. It was only through the national expansion of education and health care, by government officials rather than disparate charities and private groups, that the dysgenic could become visible and traceable in a systematic manner.

> Conditions of life … in a native reserve are such that the submerged tenth does not exist and it is probable that a discharged criminal lunatic would be received by the majority of his fellow tribesmen as an ordinary member of society.[2]

Any serious eugenic control of a population involved being able to identify and restrict individuals and groups who were of particular threat to the heritable welfare of a nation. In Kenya at this time, the administration was simply too limited and distant from the mass of the population to be able to monitor individuals regarding their eugenic status. The areas where we see eugenic thinking potentially influencing practical attitudes to social policy are those in which individuals happened to be located who were engaged with the wider debate on race and intelligence in the colony.

Oliver and 'The General Intelligence Test for Africans'

After a visit to Africa in the late 1920s, Dr Keppel of the Carnegie Corporation decided that the 'considerable sum of money' the Corporation had decided to spend on educational work in Africa should be used to fund a study on African mentality. Keppel had been struck by the unsuitability of the teaching materials used in African schools and hoped that psychological research would reveal the most appropriate materials and methods.[3] The Corporation decided to send the educational psychologist, Richard Oliver, to East Africa with the purpose of devising an intelligence test for Africans which would provide guidance on suitable educational methods and materials.

Richard Alexander Cavaye Oliver had worked with two of Britain and America's most eminent experts on intelligence testing. Oliver had been a highly successful student of the psychologist Professor Godfrey Thomson at Edinburgh University and had then won a Commonwealth Fellowship to study at Stanford University, California, with Professor Lewis Terman. Terman was well known for devising a new version of the Binet–Simon test, the Stanford–Binet, which became 'the standard for virtually all 'IQ tests that followed';[4] he was responsible for the introduction of the term 'IQ' into the language.[5] In his early publications, like his 1916 book, *The Measurement of Intelligence*, Terman emphasised the innate and immutable nature of intelligence and wrote of the need to curtail the fertility of high-grade defectives, located by intelligence testing. However, in his later work, as Gould points out, Terman became more cautious in his attribution of intelligence purely to heredity and placed more stress on environmental factors.

Godfrey Thomson, who along with Charles Spearman developed the mathematical technique of factor analysis,[6] formulated intelligence tests primarily to isolate intelligent children from poor families unable to afford education. Thomson, who was later knighted for his work, became Professor of Education in 1924 at Moray House College, the teacher training college within Edinburgh University.[7] Thomson believed that heredity was a powerful factor in intelligence: in the 1946 Galton Lecture, organised by the Eugenics Society, he argued that national intelligence was declining because of the differential birth rate. He was particularly concerned by the education system for girls, which he believed sifted out the intelligent to 'destroy their posterity'.[8] It was at Thomson's suggestion that Keppel selected Oliver to do the research for Carnegie.[9]

Oliver's academic credentials and training in the area of educational psychology and intelligence testing were of the highest internationally. Unlike many of the other figures in the Kenyan discourse on brain and intelligence, such as Gordon, Sequeira and Vint, Oliver was a highly qualified expert on the subject. After he left Kenya, Oliver went on to become Professor of Education at Manchester University. In his later work, Oliver showed less interest in the innate, hereditary nature of intelligence. He made a plea for research into broader forms of aptitude tests, and while still appreciating the value of intelligence testing, he also pointed out its limitations, arguing that it could only ever test what has been learnt.[10]

Oliver arrived in Kenya in March 1930. He was based at the Jeanes School, and thus one of the purposes of his work was to find measures that would indicate which pupils were suitable for training at the school. Jeanes schools had been established in southern states of the United States for African-Americans. They aimed to improve entire communities through the education of adults who became Jeanes teachers and were then placed in villages to pass on their knowledge through schools and influencing village life more generally.[11] The Jeanes School in Kenya was established in 1926 in Kabete, not far from Nairobi, with the help of a donation of $7,500 from the Carnegie Corporation.[12] The Carnegie Corporation continued to support the school financially; two-thirds of the salaries of the students who had been trained as Jeanes teachers were paid for by the Kenya government, although they were largely in the employ of missions.

The emphasis of the Jeanes School programme was on improving the general standard of living, not just on improving literacy and traditional educational objectives. Thus, Jeanes teachers were not only trained as conventional schoolteachers, they were also instructed on health education, agricultural methods and how to build better housing and 'improve village life'.[13] The wives of the trainee Jeanes teachers were also to be educated to impart instruction and advice on areas like basic midwifery, childcare and hygiene in the belief that they might also be able to influence village life for the better. The role of the Jeanes teachers seems to have been appreciated by members of the administration who were involved with native welfare in the Reserves. It was reported that apart from improving the quality of work done in the schools, Jeanes teachers were becoming increasingly effective as social workers. The Medical Department also recognised their value by suggesting that the work of the Jeanes teachers be combined with that of the medical dispensers in the Reserves.[14] The Jeanes School steadily grew in size, with ninety-four men and women in training by the end of 1932, and the training of the teachers was becoming more specialised: of the fifty-four men in training in 1932, twenty-eight were teachers, twenty were dispensary workers and six were agricultural demonstrators.[15]

The Jeanes School represented a more progressive approach to the question of 'native development' than was conventionally found in the Kenyan administration. James Dougall, the first Principal of the school, had been a missionary and retired from the school to resume his mission work in the early 1930s. He wrote in his 1929 annual report on the school that 'the relations between staff and students are so cordial and reciprocal they may be taken to show that racial barriers largely yield to efforts of study, and sympathetic imagination and patience'.[16] In this context, the employment of Oliver should partly be understood as an attempt to apply what were seen as the most modern and progressive methods for the improvement of the education supplied:

> Engaged as we are here very largely in routine work, it is very important that we should have attached to the staff one who, by his knowledge of education as a whole and his specialist's skill in psychological studies, can keep us awake to the necessity of continually testing the results of our work and exploring more effective ways for its development.[17]

Dr Trowell, remembering his work as medical officer in East Africa, remembered the Jeanes School in a positive, progressive light, describing Dougall and his wife 'as definitely trying to get the African on ... this was quite a different atmosphere'.[18]

In spite of its progressive aspects, however, the Jeanes School failed to challenge, and in fact in many ways fulfilled, the conservative settler image of a limited African education. The approach of the Jeanes School worked in Kenya because its ethos did not arouse the fears of settlers and administrators caused by the creation of educated detribalised urban Africans: an editorial in the *EAS* about the Jeanes School commended the system for being both cheap and a form of education 'best suited to the needs of a peasant population'.[19] As Archdeacon Owen, when questioned by the Joint Select Committee on East Africa argued, the Jeanes approach was not designed to train public leaders; instead, he said that it complied with the tendency of the educational system to create artisans at the expense of higher education.[20]

Dougall himself posed the question 'Is there a fundamental disparity between Africans and ourselves in modes of thought?' in an article contributed to *Africa*, the journal of the International Institute for African Languages and Cultures (IIALC) in 1932.[21] Dougall stated that the missionary working in Africa might be expected to start off holding a belief in the universality of human nature, but often over time and through experience became convinced of differences, usually expressed in accusations of African illogicality and emotionality. The nature of African mentality was a legitimate source of interest and questioning for even the most 'pro-native' experts at this time. The interpretation of difference put forward by Dougall, however, was quite different from that of Gordon. He cited psychoanalytic literature to draw parallels between 'primitive' and child mentality psychology, reaching the conclusion that differences between African and European mentalities were caused by the emotional and cultural dictates of environment, and could be overcome through education.[22]

Given the publicity surrounding Gordon and Vint's theories, it was inevitable that Oliver's work became caught up in the wider debate on race and intelligence in Kenya in the early 1930s. The interesting point about the role of Oliver's work was that people who held quite different views on the subject took it up as evidence to defend their positions. Although the bulk of Oliver's Kenyan work took place at the Jeanes School, he also tested students at other African schools. Crucially for the debate on race and intelligence, he performed a series of comparative tests on the students of the African Alliance High School and the European Prince of Wales School, both secondary schools. This comparative study was a small part of Oliver's research in Kenya; it became the most publicised as it dovetailed with Vint's work and Oliver wrote an article about it entitled 'The Comparison of Racial Abilities' in the *EAMJ*.[23]

Throughout his research in Kenya Oliver was cautious about drawing racial implications from intelligence testing. He wrote in the Education Department Annual Report of 1930 that the purpose of his work was to select Africans for positions which required high intelligence, and that although intelligence tests might be used for making inter-tribal comparisons in intelligence:

> a caveat should perhaps be issued against their too facile use for purposes of comparing peoples of environments so different as the African and European. A mental test may properly distinguish degrees of ability among persons who have grown up with a similar cultural background; but when applied to persons from two very different environments, there is no certainty that it is measuring the same thing in each case. It may confidently be stated that in the tests used the average African performance is considerably inferior to the average European performance; but few competent psychologists would draw the inference that this indicates a corresponding inferiority in African ability. The contribution of wide cultural dissimilarity to differences in performances is an unknown quantity. The intelligence tests at present in use cannot by themselves measure the degree of difference which may exist between European and African intelligence.[24]

The stated educational purpose of Oliver's work in Kenya was not any explicit comparison of racial intelligence, but to ascertain appropriate educational methods for East Africans. In discussing Oliver's project, both the Carnegie Corporation and the Colonial Office, in the form of the Colonial Advisory Committee on African Education, made no mention of the question of race and intelligence. Indeed, shortly after this meeting, the Colonial Advisory Committee stated in another context that it was suspicious of intelligence testing that might lead to distorted claims about race.[25] The fact of the matter, however, was that regardless of the absence of explicit declarations on evaluations of racial intelligence, the implication of Oliver's research was that African educational standards were somehow different and the educational material needed to be organised accordingly. It will be argued in this discussion of Oliver's work that among those who were debating educational capacity in Kenya, the assumption of difference in educational matters was almost ubiquitous; it was over the question of the extent of this difference and its meaning for policy that disagreement arose.

In the 1930 Education Department Report, Oliver described the methods he used to devise his tests for Africans. Sixteen intelligence tests, chosen from among those most commonly used in Europe and America, were tried out on the students of the Jeanes School. At the same time, four members of the staff of the school gave their estimates of the intelligence of their students. The students' scores in the intelligence tests were compared with the estimates made by the staff. Five of the tests were found to agree sufficiently closely with the staff's combined estimates, and were therefore retained as together supplying a valid measure of intelligence. The examinee did not use spoken language in performing the tests, and the instructions could be given in English, Swahili, or his own vernacular.[26]

In the Education Report for the next year, 1931, Oliver gave an account of the progress his work was making at the Jeanes School. The tests devised by Oliver were given to each entrant to the school and the results were used by the Principal to determine the treatment of each pupil. Oliver gave a brief description of this test in this report; the test did not require writing, answers were made in the form of crosses, lines and numbers.[27] A copy of the test, unfortunately without the instructions, is in the Kenya National Archives.[28] Oliver gives some description of the test in his article in the *EAMJ*:

> its component tests are non-verbal, consisting of problems dealing with pictures, numbers, letters and other symbols. The pupils are not required to write words, but only to make crosses, letters, numbers or other signs on the paper. In one of the tests, for example, the problem is to make a cross through any part of the paper which is absurd.[29]

The aim was to make the intelligence test widely available to education-ists as an aid to admitting pupils, for classifying them and vocational guidance. Copies of the test, with instructions and the marking system, were printed and given to the education department to enable the wide-spread use of the test by other schools. The Jeanes School was committed to using the 'General Intelligence Test for Africans' after Oliver left the country at the end of 1932; it was felt that his work at the Jeanes School had 'left us a legacy of permanent value'.[30] Oliver tested all the Jeanes teachers in the field and those in training, and the test was used to select candidates for training in dispensary work.

While in Kenya, Oliver also performed tests of musical ability; his subjects were ninety pupils at the Alliance High School. Oliver claimed to have discovered that in comparison with American children of approx-imately the same school standing, the Africans were superior to the Americans in their sense of intensity (loudness), time and rhythm, in that order. They were inferior in the senses of pitch and harmony and in the memory of tunes, in order of increasing inferiority. Oliver argued that these comparative strengths and weaknesses were related to the qualities found in African music. He intended to make the next stage in this study a comparison of the various East African tribes in respect of musical talent.[31] After Oliver left Kenya, Mr. W. H. Taylor continued this work on musical testing at the Jeanes School, claiming to find differences in musical talent between tribal groups which correlated with tribal IQ test results.[32]

Oliver and Taylor's examination of musical abilities to draw compar-isons between different Kenyan tribes and African and European musi-cality was part of a package of analysis undertaken to guide European education administrators. It seems that the testing that was undertaken under the auspices of the Jeanes School was not motivated by a desire to attempt to assert African inferiority; it was a more generalised desire to categorise and rate African mentality. Although Oliver's research had many similarities with the arguments made by the likes of Gordon and Vint, it will be demonstrated that it did not sit easily with the rhetoric of their campaign. Oliver's work was intended to guide educational methods at an immediate and pragmatic level at places like the Jeanes School. Intelligence testing was a method used by Oliver to enable him to devise new methods and materials for the teaching of reading, including the compilation of a basic reading vocabulary in Kikuyu.[33] Where the campaigning of Gordon and his supporters envisaged a grand scheme of research with imperial dimensions, Oliver was working at a more grass-roots and less politically ambitious level. Beneath the more tempered language used by Oliver, however, it will be shown that there was the same assumption of biological difference in intelligence.

[153]

At the beginning of his time in Kenya, Oliver had been unwilling to use intelligence testing to make racial comparisons. However, his comparative testing of African and European high school students led him more deeply into the debate than he seems to have originally intended – this experiment was performed at the request of Scott, who was one of the colony's more vocal doubters of Gordon's insistence on innate mental difference. In 1932 Oliver published an article in two parts in the *EAMJ*, presenting his findings. Oliver began his article with a definition of the abilities of the races he was examining; he defined these as 'culture-making abilities', in which the chief factor was general intelligence, 'a factor entering into the performance of every kind of task'.[34] Oliver began by discussing 'comparison of cultural achievement' as a measure of racial difference. Although accepting the notion that the achievements of European culture were superior, Oliver dismissed the idea that those cultural achievements could necessarily indicate inherited biological superiority. He argued that there were external factors in cultural advances such as accident, physical environment, ethnographic environment (proximity to other groups from whom ideas could be copied), size of population and the existing culture base.

The second comparative tool described by Oliver for measuring racial abilities was psychological tests, most importantly in the form of intelligence testing. In his discussion on psychological testing, Oliver introduced the readers of the *EAMJ* to American 'findings' on race and intelligence testing and introduced his own results from working in Kenya. Oliver's research was greatly influenced by the intelligence tests that had been made on African-Americans in the United States, tests that were more favourable to whites. In reference to the intelligence testing in America, Oliver remarked that: 'Almost without exception, the average intelligence of the negro groups has been found to be definitely lower than that of white groups.'[35] Oliver reported that the average black American result was 20 per cent below that of the average white American result.

The 'General Intelligence Test for Africans' was performed on 124 European boys at the Prince of Wales Secondary School, and the average age of these schoolboys was just under fifteen. The number of African schoolboys tested was ninety-three, they were attending the Alliance High School, and their average age was estimated to be nineteen. Oliver claimed to have eradicated the effects of distorting factors caused by environmental factors such as language and cultural difference as much as possible, but he was forced to admit:

It is probably safe to say that no test has yet to be devised which draws equally upon the experience of advanced and primitive groups, and which is therefore a completely true measure of the intelligence of such groups.

> Applied to groups differing widely in culture, the tests are measures of intelligence plus a cultural factor of unknown amount.[36]

Oliver admitted that the quality and consistency of the schooling received by the pupils at the Prince of Wales School were better than at the Alliance High School; the cultural and linguistic differences of the two groups were not specifically acknowledged. The average European score was 312 points, and the average African score was 266. Thus, Oliver concluded that African mental development reached 85 per cent of European mental development, and that about 14 per cent of Africans reached or exceeded the European average; Oliver pointed out the similarity of his figures with the results of the testing carried out in the United States.

In the concluding part of the article, Oliver also used the similarity between his and Vint's figures (Vint's study on the cell content of the prefrontal cortex claimed to indicate that African development was at 84 per cent of European[37]) as a confirmation of the validity of each experiment. In his discussion of head and brain measurements as a method for comparing the abilities of race, Oliver questioned, but did not altogether dismiss, Vint's correlation of brain size and weight. He more unequivocally applauded Vint's histological work on cell content:

> The use of brain measurements as a basis for the comparison of racial abilities has recently been extended in an interesting and promising direction. Dr F. W. Vint … clearly, and I think rightly, regards his findings as having some bearing on the comparative mental development of East Africans.[38]

In this article, then, Oliver clearly placed his work in the same sphere as Vint's in terms of the results obtained and their meaning for racial comparison. There was one difference in Oliver's results when compared with Vint's research on cell content: although they agreed that the average African standard was in the region of 85 per cent of the European standard, there was a divergence in the percentage of Africans said to reach or exceed the European average. Vint said that only 6 per cent of the East African brains that he performed his post-mortem analysis on reached or exceeded the European average, whereas Oliver put the figure at about 14 per cent for his examinees.[39]

In another article, published in *Africa*, Oliver again emphasised the preliminary nature of his research when it came to differences between Europeans and Africans, although he did believe that the results he obtained 'give a first approximation to the truth'.[40] He suggested that in educational terms, 'the average European is educable enough to pass through the primary school and a few years of secondary school', while the average African 'would be capable of absorbing the education provided in the European primary school'. He emphasised that both European and

African intelligence varied over a wide range, but the African 'distribution would be displaced somewhat toward the lower end of the scale, but there would be considerable overlapping with the European distribution'. He believed 'a small percentage' of Africans would be capable of undergoing a university education.[41] In the following issue of the same journal, Oliver's 'General Intelligence Test for Africans: A Manual for Direction' was positively reviewed as 'a useful beginning in the important subject of the measurement of intelligence among Africans'.[42]

Scott, who requested that Oliver perform the comparative tests on European and African students, took the 14 per cent figure as crucial in the debate on intelligence. In the issue of the *EAMJ* before the one in which Oliver published his work, Scott wrote an article on educable capacity in which he quoted Oliver's findings, to which, as Director of Education, he had already had access. Scott's article was a critique of Gordon and Vint's campaign. Although he acknowledged the similarities in Oliver's and Vint's conclusions, Scott took this 14 per cent figure as highly significant:

> 14% of the Africans reached or exceeded the average score of Europeans and 10% of the Europeans fell below the average of the Africans. If we could conclude that 14% of all Africans are above the average of all the Europeans, the figure of 80,000 which I gave above becomes about 400,000. Anyone who would draw such a conclusion on one test would indeed be very rash but the evidence as far as it goes does seem to indicate the existence of a large number of Africans who, given good environmental conditions and fair training, are of relatively high educable capacity.[43]

The notion of inherent racial difference was so entrenched that the argument of a relatively progressive expert such as Scott, that intelligence testing at least demonstrated that African intelligence was not as low as previously thought, was relatively radical. Oliver was working from the same assumption of difference in intelligence between races. The fact that his work took place under the auspices of the Jeanes School demonstrated that belief in mental difference was ingrained in even the more moderate institutions in Kenya.

Juvenile delinquency and Kabete Reformatory

The treatment of juvenile delinquency in the discourse on African development provides an insight into the role of eugenic thinking in social policy in Kenya. Anderson has pointed out that for certain sections of the European settler community 'law and order had been a near-obsession', making it a constant political issue from an early stage in the colony's history.[44] The issue of juvenile crime fed into these fears about the criminal threat against the European minority and juvenile delinquency became a focal point for debates about the deleterious effects of urbanisation and 'native development' on African behaviour. This section will show how the African juvenile delinquent became a quintessential figure in the Kenyan discourse on development and social change, and how eugenic and progressive social policies on juvenile delinquency overlapped.

During the 1930s, there was a movement in Kenya to adopt recent British legislation,[45] which was generally in tune with modern British social policy. Although the changes made to the legislation on juvenile delinquency conformed to this wider legal trend, the process by which these changes were judged and evaluated as appropriate to the 'Kenya native' uncovers a fascinating discourse on African social policy. Juvenile offenders were a concern for the colonial government from an early stage, but it was not until the 1930s that modern reappraisals of delinquency occurred following a rise in the number of juvenile convictions and a subsequent panic about the effects of urban conditions on African youth. The role of biological thought and urbanisation will be examined as recurrent themes in the Kenyan analysis of juveniles, in particular the juveniles of Nairobi who seemed to represent a generation of detribalised and thus unmanageable natives. These anxieties arising from urbanisation were closely linked to the social concerns broached by eugenics in Kenya.[46]

The management of juvenile offenders underwent a transformation in the first half of the 1930s when juvenile crime, the social problems that caused it, and the best methods of treating it became subject to intense analysis in Kenya. This led to a series of legal changes and a restructuring of the reformatory system along the lines of the most recent British developments. In this process, four important reports were produced; two ordinances were passed relating to juvenile offenders in 1933 and 1934; and two further amendment acts were passed in 1935 and 1936.[47]

The treatment of children and adolescents was confused by racial prejudices and the financial limits inherent in the colonial state at that time. Compulsory education for European children in Kenya was introduced in 1934. This came about after a long-running debate in the press and Legislative Council about the risk of a poor white class developing.[48] The introduction of universal education was a prerequisite in Britain for bringing about changing attitudes towards childhood. In Kenya, the desire to maintain white prestige as well as the financial constraints on the colonial state meant that childhood was a luxury that was unevenly distributed between the races. In 1932 there were 578 boys and 149 girls attending school in Nairobi, and a further 100 pupils attending night school. This was out of an estimated population of 2,608 juveniles under the age of sixteen living in Nairobi, with another 838 resident in the Native Location.[49] Despite the recommendations of the *Report on Juvenile Crime* in 1934 that African education be made compulsory in Nairobi,[50] the financial expenditure was not forthcoming.

Race undoubtedly played a role in the analysis of the causes of juvenile offending, and therefore its treatment. The fact that there was a majority of Kikuyu among the inmates of Kabete Reformatory was attributed by the Superintendent, Captain Wood, to race:

> Racial turpitude is in my opinion the only answer, coupled with the undoubted fact that there is more unrest amongst this large, but perhaps not the largest, branch of the native population.[51]

It is not surprising that the problems of juvenile delinquency in Kenya were discussed with reference to an inherent incapacity of the African brain to deal with the complexities of modern, urban life. This was a theory put forward by Dr Gordon in his discussions about dementia praecox, in which he argued that the effects of education and 'Europeanisation' were liable to stress the native brain beyond its capacity, resulting in mental breakdown. Gordon drew a parallel with adolescent European schoolboys, who, according to the theory, sometimes suffered from psychosis induced by excessively advanced academic activities forced on an immature brain. Captain Wood wrote:

> It would appear to be quite necessary to determine what percentage of inmates are unfit for town life, either by reason of mental inefficiency, or criminal tendencies; also to devise some means of keeping boys out of towns as practically all reconvictions occur in townships.[52]

This attitude to juveniles in Kenya can only have been increased by the research carried out by Dr Gordon at Kabete Reformatory.

The process of seriously examining the problem of juvenile delinquents began, therefore, in 1930, when Dr Gordon conducted his controversial research on the intelligence of the inmates of Kabete, writing a 'Report of a Survey of the Inmates of Kabete Reformatory for the Purpose of Detecting Amentia (Mental Deficiency)'.[53] Gordon's research was undertaken privately, but the report was circulated among the officials and settlers interested in Kabete. As well as being sent to the British Eugenics Society, Gordon's report was also, via the Director of Medical Services, distributed within the Kenya government.

Gordon's report on the Reformatory was an important statement of his methods and ideas. Its significance here is in how it reflected and possibly shaped ideas about dealing with juvenile offenders in Kenya; or, as Gordon's report indicated, not dealing with them. A summary of his conclusion was printed in the Annual Report for Kabete of 1930.[54] The 'finding' that only 2.7 per cent of the population of the Reformatory were classified as 'low grade normal', 11 per cent were 'borderline normal', and 86.3 per cent were aments, and that not one of the inmates tested was found to reach the standard of 'average normal' by European standards, was clearly rather shocking. As the Chief Justice indicated, it brought up issues about the possibility of the institution functioning with any correctional, educational purpose:

> I am directed by His Honour the Chief Justice to state that the most alarming part of the Report [for Kabete, 1930] is that dealing with Dr Gordon's examination of the mentality of the inmates. It is amazing that 2.7 (6 boys) only can be placed in the 'normal' category and even they are 'low grade'. Those facts appear to militate against any permanent reformation of the great majority of the inmates being possible.[55]

The Minutes of the Committee of Visitors for Kabete Reformatory show an interest in Gordon's analysis but no full discussion of opinions on the report is recorded in the surviving document.[56] Gordon's expertise in diagnosing mental problems among African juveniles was clearly accepted: in his 1934 contract of employment as Visiting Physician at Mathari Mental Hospital, Gordon was also required to visit Kabete to report and advise on the mental health of youths brought before a juvenile court, as well as assess the mental health of prisoners at Nairobi Prison.[57]

It is unfortunate that the opinions of the committee members are not documented as the committee included the Chief Justice, the Chief Native Commissioner, the Director of Public Works, the Senior Resident Magistrate and the District Commissioner. The Colonial Secretary, however, did forward the Chief Justice's comments, quoted above, to Dr Gilks. Gilks' response was to reiterate the qualification issued by Gordon in the report: that it was to be expected to find fewer people of normal capacity in a Reformatory and that the report used classifications based on European standards. He further attempted to soften the extremity of the report by saying that a re-survey might transfer some of the high-grade aments to low-grade of borderline normals, and that also pending re-survey, the borderline normals should be regarded as normals, bringing the total figure for normals up to 13.7 per cent.[58] In this letter Gilks also suggested that some sort of separate institution for segregation was advisable:

> Those left in the high grade class after re-survey could be divided into those suited for some suitable training or for socialization after discharge and those for whom permanent segregation is necessary. Medium grade 'aments' are suitable only for permanent segregation.[59]

The eugenic agenda behind the advocation of segregation at this time adds an extra dimension to Dr Gordon's findings. This dimension is strengthened by Gilks' elaboration on amentia at the end of his letter:

> It is not possible to overlook the fact stated on page ten of Dr Gordon's report that mental deficiency of all degrees is the result of defects and deficiencies in certain areas of the brain and incapable of material amelioration by treatment or education. It is also important to note that evidence is accumulating from research elsewhere that the brains of backward races exhibit certain racial inborn defects and deficiencies which necessarily limit the level of mental development and behaviour to which they can attain.[60]

The problem that eugenicists like Gordon and Gilks faced was that any change in policy by the Kenya government in the treatment of juvenile offenders was to be dictated by metropolitan thinking which was increasingly emphasising the importance of reform and rehabilitation through the Borstal system. Segregation, rather than being on a eugenic basis, entailed isolating the hardened recidivists to prevent them from corrupting those who might reform. This should not be taken to imply that Gordon's line of thinking had no resonance with British policy, as can be seen in report on 'Juvenile Welfare in the Colonies' of 1942. This was the report of the Juvenile Delinquency Sub-Committee of the Colonial Penal Administration Committee, whose terms of reference were to consider the prevention and treatment of juvenile delinquency in British colonies. The following was commented:

> The percentage of high-grade defectives or mentally retarded children in the total figures for juvenile delinquency is notoriously high. This is an aspect of delinquency that does not appear to have been investigated fully under colonial conditions, and it is probable that many children who have been either whipped or imprisoned for want of alternative methods of treatment, were in fact psychologically abnormal in some way or other, and therefore unlikely to respond to either treatment. In this country, medical or educational psychologists are more and more frequently consulted about the welfare and treatment of a delinquent child ... Either of the two last officers [medical and educational psychologists] should be given opportunities to engage in research, for it is unlikely that further progress will be made in the practical application of psychology until considerably more is known about the normal and abnormal psychology of colonial peoples.[61]

The effects of Dr Gordon's well-publicised study were compounded by the fact that in the year 1930 there was a substantial rise in the number of convictions of juveniles in Nairobi, largely because of the increased poverty caused by economic depression. In 1929 there were 154 juvenile convictions in Nairobi; in 1930 there were 322 and this continued to rise to 463 convictions in 1931.[62] Despite the correlation of the rise in juvenile crime with the increased poverty caused by the economic slump, the debate concerning juveniles centred on the anxiety aroused by the appearance of an urban African youth rather than on the effects of poverty.

In the early 1930s, concern about African juvenile delinquency in urban areas reached a point where official intervention was required, and a further step was made to define and acknowledge the native child. Until the early 1930s, child crime had been dealt with by Kabete Farm/Reformatory and the odd burst of outrage in the press. But in 1932 the *Crime Committee Report* was commissioned by the Kenya government.[63] Its purpose was to make recommendations on how to deal with 'large numbers of potential juvenile offenders, habitual criminals, vagrants and unemployed in Nairobi and the areas bordering on Nairobi'.[64] It therefore considered juvenile crime as a part of the social problems afflicting Nairobi, and as such was broad in its examination of the conditions that contributed to delinquency. The Chair was Armigel Wade (at that time Chief Native Commissioner). The committee also included a regular on such committees, Captain H. F. Ward, who was European Elected Member for Nairobi North on the Legislative Council and, along with his wife, was a member of the KSSRI. Another member of the committee was Scott, Director of Education, who supported Dr Gordon's campaign for research into race and intelligence, although expressing dissent from its more extreme assertions about innate African intelligence.

This is not to argue that the Crime Committee had a eugenicist agenda, but it is interesting to note that some of those concerned with policy towards juveniles in Kenya at this time were aware and supportive of eugenics. The KSSRI was certainly interested in child welfare and juvenile delinquency, establishing a sub-committee on the question of children,[67] and issues such as Infant Welfare and Child Guidance were top of the list of the eugenic problems the Society proposed to discuss.[68] In a speech to the KSSRI Dr Gordon highlighted the significance of juveniles in his list of immediate needs for Kenya in the area of race improvement. Dr Gordon's fourth point, after establishing the levels of mental deficiency in all races, achieving a child cruelty act and improvement of the treatment of the mentally ill, was the need for a magistrate's court for young offenders with a psychologist and social worker attached to it.[69]

The overlap of the individuals who were interested in social policy and in eugenics in Kenya is unsurprising given that eugenics was concerned with the social implications of biological theories. It was partly the result of the small size of the Kenyan European population, but also indicates how eugenic thinking in Kenya was not just the preserve of isolated individuals, but included people who were involved in issues of social welfare and prescribing policy. There are further examples of the continuity of personnel involved in eugenics and juvenile delinquency. In 1933 Dr Paterson recommended that Dr Gordon be made responsible for the diagnosis of mental health problems at Kabete.[70] La Fontaine, whose report on delinquency will be discussed shortly, was also an active participant in the KSSRI.[71]

The *Crime Committee Report* revealed that poverty and inadequate living and working conditions were commonplace for many of the African youths in the city. Of the 3,446 juveniles in Nairobi, about a third had no parent or guardian there, about 35 per cent of these juveniles were in regular employment, and 10 per cent had access to some sort of casual work; about 800 children attended school. The infant mortality rate was high, about 150–60 per 1,000.[72] There was anxiety about the way in which the lives of children in Kenya were developing, with a drift from parental and traditional forms of authority to an anarchic and unmonitored urban existence where lack of education and employment and poor living conditions were creating a criminal and uncontrollable youth and future urban underclass.[73] This unease about the effects of development on children was primarily expressed in terms of crime and urban settlement. References to the problem of orphans or inadequately protected children in a city like Nairobi are first found in crime reports:

> Many of them have left their parents in the Reserve and have come to Nairobi to look for employment. Some of them are children of irregular and temporary unions; others are children of prostitutes and unknown fathers. In many cases these obtain intermittent employment with Indians, Somalis and others, for no regular or stated wage but for food and an occasional present. As a result of a test round-up made by the Superintendent of Police, 47 unemployed juveniles were brought to the Police Station in two and a half hours. According to the Superintendent of the Native Locations, a large number of the 838 juveniles resident in the locations spend their time wandering about Nairobi looking for work. A considerable number of them have no relatives in the locations, and at night are compelled to search for somewhere to sleep.[74]

Juvenile crime in Kenya partly developed as a result of poverty and the dislocation of families, largely due to employment practices and the failure of the colonial state to invest in the mechanisms for providing welfare networks that had proved necessary with rapid urbanisation in Europe. Juveniles became a problem only when they were perceived as a threat to order through criminality, but the failings of colonial government that the problem of juvenile delinquency uncovered were fundamental. The level of welfare provision required to cater for the social problems that lay behind delinquency in a modern urban environment required expenditure and a concomitant rethinking of policy that was incompatible with the colonial venture.

As well as pointing out the social problems in Nairobi, the committee also gave voice to the theories of innate physical or mental inadequacy as a cause of juvenile delinquency:

> Among juvenile delinquents in London it has been found that physical defects are 1.25 times as frequent as among non-delinquents from the same areas ('The Young Delinquent' by Cyril Burt) … In this connexion I would stress the desirability of a comprehensive medical examination, both physical and mental, of every juvenile delinquent. Defects might be found which, possibly, would throw light on the causes leading up to the lapse and which might modify considerably the punishment afflicted.[75]

The tendency to attribute such social problems to individual physical or mental failings had a history in the discourse of the urban poor in Britain. Juvenile delinquency was the kind of social question to which the British Eugenics Society devoted much of its analysis.[76] It is interesting that the *Crime Committee Report* cited Burt's work on juvenile delinquents. Here, Burt emphasised the multiplicity of both contributing factors and major causes of juvenile crime.[77] He summarised the causal conditions into four categories: hereditary, environmental, physical and psychological, and asserted that heredity was the major factor in delinquency in 36 per cent of boys and 41 per cent of girls.[78] An ambivalence towards juvenile criminality is clearly indicated in the *Crime Committee Report*; it analysed the environmental causes of delinquency, but was reluctant to eschew the possibility of biological defect as a cause. It was the concept of the unsuitability of urban life for the African that helped to resolve this ambivalence.

In the light of the report it was decided to commission an official investigation into how to manage actual offenders. Thus in 1932 La Fontaine, Provincial Commissioner, was sent to Britain to investigate the Borstal system of handling juvenile criminals and to report back with his recommendations.[79] La Fontaine was involved in the KSSRI, and had held a prominent role in the first public meeting of the group, described as being 'awfully good in support' to Dr Gordon.[80] By this time Acting Chief Native Commissioner, La Fontaine wrote a 'Report on the Applicability to Kenya of Methods Pursued in Borstal and Other Reformatory Schools in England' in which he addressed the question whether 'African human nature' was amenable to a reformative policy.[81] The report concluded that Borstal methods could effectively be applied to Kenya. Indeed, La Fontaine argued that their suggestible mentality made Africans easier to rehabilitate:

> officers with close personal acquaintance of the African will agree that he

possesses in undeveloped form many of the qualities which go to make up of the average European lad, and that where he lacks these qualities, his plastic nature is such that they can be grafted on to him provided that good influence is continuously and intelligently exerted ... there are good grounds for believing that in the case of the more malleable African results should be even more gratifying.[82]

The issue of mental capacity was also brought up by La Fontaine, necessarily after Dr Gordon's report on the intelligence of the inmates at Kabete, which as shown above was interpreted as meaning that the average Kenyan juvenile criminal was, owing to mental deficiency, incapable of reformation:

> Critics will say with some plausibility that the African mentality is too low, his moral horizon too limited, for him to respond to improving influences, if pitched on too high a plane. Our hope lies in the natural trust of the African boy or man in the European, his natural respect for the stronger and better, and his genius for imitation: and I believe our hope will not be disappointed if we advance boldly, only tempering our boldness with the caution necessitated by a knowledge of African character and African conditions.[83]

La Fontaine recommended that Dr Gordon perform a 'psycho analytic' [sic] examination of the inmates.[84]

The effect of La Fontaine's recommendations was rapid. In 1933 the Juvenile Offenders Ordinance was passed in Kenya. This Act defined a child as being under fourteen, and a young person as being under sixteen, and that children should have separate court hearings, and that children and young people should be protected by the Commissioner of Police while being detained to prevent them from associating with adults (other than relatives) who had been charged with an offence. It was also decided that court procedure should be made more appropriate to children and young people. If a sentence of imprisonment was passed there were to be industrial schools and reformatories, and the sentence should be no less than three years and no more than seven.[85] Thus children and adolescents were recognised as needing particular protection within the legal system.

In August, a committee was set up to investigate the success of the 1933 Juvenile Offenders Bill. The *Report of the Committee on Juvenile Crime and Kabete Reformatory* was not produced until 1934, by which time the bill had been passed into law. The Chairman of the committee responsible for this report was La Fontaine; its other members were Scott, Willcocks (Commissioner for Prisons) and Captain Ward. The committee's terms of reference were to consider possible measures to deal with juvenile crime and to make recommendations about the future of Kabete Reformatory, bearing in mind that no measure could be entertained involving any considerable expense in the immediate future. Among other recommendations, the report suggested that compulsory education be provided for native children in Nairobi, that Probation Officers be introduced and that new classifications of detention centres should be instituted along British lines, with an Industrial School for those who had not committed any serious crime but were vulnerable to criminal influences through problems such as vagrancy.[86] The report made it clear that these different institutions should be grouped together under the title of Approved Schools, and run according to a unified policy.

Thus Kabete, according to the new classifications recommended, would change from being a Reformatory to a Training School (for boys aged fourteen to nineteen who had committed a crime), along the lines of Borstal in England. The committee also recommended another reformatory school be established at the government station at Eldama Ravine. The concern with the reclassification of institutions for juveniles was influenced by the English approach, seen in the 1933 Children and Young Persons Act, which deliberately blurred the distinction somewhat between reformatories and industrial schools by placing them under the grouping of Approved Schools. This change reflected the strengthening of the idea that children who were neglected and suffering from environmental deprivations were more likely to offend. The *Report of the Committee on Juvenile Crime and Kabete Reformatory* caught the spirit of this new sentiment:

> As to the training at these institutions, we agree that the first essential in the 'reformation' of the African child or youth, who shows signs of criminal tendencies, is to be taught discipline, and, in this, the ideal to be aimed at in Kenya, should follow that of England and America, viz. That discipline should be instilled, not by fear of punishment, but rather by careful leadership and example. We also agree that this can best be obtained by hard work, hard play, the fostering of the community spirit and by as much freedom as possible.[87]

In 1934, influenced by the *Report of the Committee on Juvenile Crime and Kabete Reformatory*, the 1933 Juvenile Offenders Ordinance and the Reformatory Schools Ordinance were repealed and a new Juveniles Ordinance was passed in Kenya, which was based on recent English legislation. Most of the provisions of the Ordinance of 1933 were re-enacted, except that the age of a 'young person' was redefined as being between fourteen and eighteen, rather than sixteen; remand homes were to be established in every district; and Approved Schools under the supervision of an Approved School Board were to replace Industrial and Reformatory Schools. These new Approved Schools were to be divided into three classes: Class I for the detention and training of unconvicted children and young persons found to be living in an unsatisfactory environment where they might be exposed to criminal influences; Class II for the committal and training of convicted children up to the age of sixteen; and Class III, run according to the Borstal system, for the committal and training of convicted young people up to the age of eighteen.[88] This new piece of legislation was then amended the following year, in 1935. The 1934 Ordinance provided for the committal of a juvenile to an Approved School for a period of from three to seven years, but unless the juvenile had attained the age of seventeen at the time of the committal, he could only be detained in a Class III Approved School until the age of eighteen. The amendment of 1935 made it possible to detain a young person in a Class III Approved School until the age of twenty-one. According to the 1934 law, a juvenile who had escaped from an Approved School could only be punished by increasing the period of detention by six months at the most. This was considered to have little effect on the culprit because the punishment only took place in the distant future and did not set an adequate example to fellow inmates. An amendment was introduced to allow whipping as a punishment for escape.[89] Yet another Juvenile (Amendment) Ordinance was introduced in 1936: the Chief Inspector of Approved Schools had complained that the disciplinary measures which it was permissible to inflict under the Juveniles Ordinance were insufficient for incorrigible inmates, and that such inmates were also a harmful influence on other juveniles in the schools. The amending Ordinance, therefore, empowered the Governor to commute the sentence (or remaining part of it) of a juvenile to a sentence of imprisonment in the Nairobi Prison (this was in conformity with British legislation).[90]

In the 1930s, therefore, there was a major rethinking of policy towards the treatment of juvenile offenders in Kenya, characterised by attempts to bring the running of the reformatories up-to-date with the British system. In January 1935 an Officer from the Borstal Service in England took charge of Kabete Approved School.[91] Commander W. H. L. Harrison, who had been housemaster at HM Borstal Institution at Portland, and before that a commissioned officer in the Navy, became the new superintendent.[92] According to the 1934 Ordinance, Kabete Reformatory was gazetted as an Approved School, which combined the work of Class II (up to the age of sixteen) and Class III (up to the age of twenty-one, according to the amendment of 1935) schools. By about 1936, the Approved School at Kabete had accommodation for 150 juveniles, but only contained 94.

In 1937 a further Class II Approved School was established at Dagoretti on fifty acres of land purchased from the Council of the Alliance of Missions. The school at Dagoretti was to be run in close liaison with the Kabete School and when an inmate at Dagoretti reached the age of sixteen he was transferred to Kabete.[93] Dagoretti, like Kabete under the new Approved School system, taught agriculture rather than trades in order to encourage ex-inmates to stay out of towns. At about the same time, a juvenile ward in Nairobi Prison was gazetted for inmates of the category Class III, who were considered unable to benefit from being in an educational as opposed to a prison environment.[94] The only established Class I Approved School was the Salvation Army Welfare Centre in Nairobi. It had been recommended by the 1934 Committee on Juvenile Crime that a Class I Approved School be built at Tigoni and run by the Roman Catholic Mission, but it was considered too expensive and so the existing Salvation Army facilities were used.[95]

The methods used to instil discipline in the new approved schools were in keeping with the ideals behind the Borstal system, avoiding brutalising punishments and emphasising the value of leadership and example, rewarding good behaviour with greater freedom and trust, and encouraging hard work:

> The first essential of any reformative institution is discipline. This should not conjure up visions of chain gangs, bolts and bars, and warders armed with rifles. The degree of discipline should be felt rather than seen ... At Kabete boys can be seen working in small parties under no supervision beyond that of an occasional visit by one of the Staff. Naturally, at first, trust in a boy is sometimes misplaced, but often after one warning and no punishment the boy is found to be worthy of the trust placed in him.[96]

The approach adopted after the reforms of the 1930s was characterised by an optimism and faith in modern methods that emphasised the environmental causes of juvenile crime and the capacity for rehabilitation through the right reforming environment. This can be seen in the report written for Hailey by the Superintendent of Kabete, Harrison, who had taken over at Kabete in 1935: 'Juvenile crime will always be a source of concern to the authorities and it will always be with us; but by applying modern methods of reform to African juvenile delinquents it is believed that a great deal can be done to turn potential criminals into useful citizens.'[97] The environmentalism that lay behind the new metropolitan thinking introduced into the treatment of juveniles in Kenya was, however, given a racial twist. This can be seen in Willcocks' (Commissioner of Prisons and Chief Inspector of Approved Schools) report to Hailey's commission in reference to the new Borstal system:

> In the opinion of the writer the average African juvenile is extremely receptive of influences. He is an imitator, a hero-worshipper and plastic stuff for the moulding; and it is for these reasons that environment during his formative years must play a large part in shaping his future conduct of living.[98]

According to this line of thought, the particularly malleable nature of the African youth was endangered by the conditions of urban living:

> The Approved Schools may be compared with the last reserves to be used in the battle against juvenile crime, in which better parental control and improved housing and hygienic conditions should be the first lines of skirmishes and compulsory education the main attack. If the latter fail in their object the training in the Approved Schools may be regarded as the last hope before the juvenile delinquent becomes the adult criminal. A skilful use of reserves has turned the scale in many a battle. If the Approved Schools are well administered and their aim is clear there should be an equal chance of success.[99]

Whereas under the old reformatory system inmates were taught an artisan's trade that might encourage them to seek employment in a town, the Approved School system returned to an agricultural training. Echoing the emphasis of the original Kabete Experimental Farm, the main training was to be in animal husbandry, gardening and agriculture. This change in policy was a deliberate effort to push the ex-inmates into rural areas on discharge, either in Reserves or in European employment.[100]

The approach towards the treatment of juvenile offenders in Kenya underwent a rapid change in the first half of the 1930s. There developed a pronounced official interest in researching and analysing juvenile offenders. There were three causes for the timing of this explosion of interest. The first had its origins in metropolitan intervention and the broad legal reforms encouraged by government; Kenya's penal code was being rewritten in the early 1930s, hence, as Morris and Read have argued, it was a period in which crime and punishment were being closely scrutinised.

The second factor that may have induced this interest in juvenile delinquency was a local crisis in Kenya's prisons. Anderson has shown that in the 1930s there was a substantial increase in the total number of convictions in Kenya's subordinate courts (29,783 in 1929, rising to a peak of 50,465 in 1934). Increasing proportions of convicts were also going to prison because of the inability of many non-custodial convicts to pay their fines. Hence, enormous pressures were being placed on Kenya's prisons.[101] Overcrowding increased the likelihood of juveniles mixing with more hardened adult convicts. The crisis in Kenya's prisons, combined with this dramatic rise in crime figures, encouraged a rethinking of the punitive and rehabilitative facilities for juvenile offenders.

The third influence that may have contributed to this interest in juveniles was that of Kenya's eugenicists. Gordon provoked official interest in Kabete in 1930, and two of the individuals consistently involved in the management of juvenile offenders were associated with the KSSRI – La

Fontaine and Captain Ward. The eugenicists stirred up interest in wider issues that had implications for juvenile delinquency, such as African development and reformability. The fact that the starting point of Kenyan eugenics on juvenile education and reform was pessimistic, while still supporting the introduction of progressive methods that sought rehabilitation, captures the ambiguous stance of the KSSRI on the question of native development.

The recurrent theme connecting the changing approaches to juvenile delinquency was doubt about the suitability of the African for urban life. This can also be seen later, in the Report of the Committee on Young Persons and Children published in 1953, and chaired by Humphrey Slade (Gordon's son-in-law): 'In the case of Africans we have found that the impact of western civilisation is steadily breaking down the pre-existing influence of parent and clan.'[102] The juvenile delinquent in Kenya came to represent the problems of urban poverty and social breakdown induced by developments imposed by the colonial state. However, the acknowledgement of the environmental factors in the rise in juvenile crime was tempered by the discourse on either the neurological or emotional incapacity of the African to deal with change. The 1930s saw a transition from the blatantly eugenic racism of Dr Gordon to a more environmentalist approach, but one that still maintained an emphasis on the problem of African incapacity.

Mathari Mental Hospital, criminal lunacy and criminal responsibility

It is unnecessary to go into great detail on the legal history of insanity and the institutional history of Mathari Mental Hospital; they are the subject of the work of Gail Beuschel, who has provided a valuable contribution to the history of medicine and madness in colonial Kenya.[103] This section will examine the implications of the theories about African mental incapacity for the idea of criminal responsibility and the interaction between Gordon's eugenic theories and his work as Visiting Physician at Mathari Mental Hospital, a position which, until 1937, endowed him, as Kenya's only expert in the field, with considerable influence within the sphere of mental health.

Gordon's approach to African mental disorder, as well as mental deficiency, was, as he revealed in his article in a major British psychiatric journal, distinctly biological.[104] The idea that the East African native was liable to adolescent breakdown under the 'impact of civilisation' owing to an under-equipped frontal brain was developed in an article in 1936. Gordon found that of his African adolescent patients at Mathari about whom reliable information could be obtained, all were educated; he took this as evidence of the danger of exerting too much intellectual pressure on a deficient prefrontal brain.

> I prefer to think of it as a series of pressures; pressure of the developed upon the undeveloped mind; pressures of thought, emotion, behaviour; of intellect, imagination, foresight; of suggestion and example; of venerable codes, customs, and national ideals, founded on a religion of truth, justice, and love – all forming an unprecedented experience for the Native brain involving inevitable adjustment or inevitable catastrophe.[105]

Vaughan has demonstrated a similar emphasis in attitudes to African insanity in her study of Zomba Lunatic Asylum in Nyasaland, where insanity among educated Africans, whose 'idioms of madness' were European, aroused disproportionate concern.[106]

The diagnosis of a high incidence of amentia and mental instability in the African population had difficult legal implications when it came to defining criminal insanity and responsibility. This can be seen in the memorandum written by Dr John Carman, a Medical Officer who had temporarily been posted at Mathari in 1937 after Gordon had left and before the arrival of the first qualified psychiatrist at Mathari, Cobb, whose tenure was brief owing to his own psychological problems and scandalously unprofessional practices.[107] Carman, who had also been posted as Medical Officer in charge of Nairobi Prison, was requested by the Governor to write a memorandum on the medical aspects of trying Africans. The question of examining the psychological state of Africans charged with capital crimes was of particular pertinence after a recent case in which a slight delay allowed for the last-minute discovery of medical evidence that prevented an execution taking place.[108]

One of the problems raised by Carman was cases of violent crime, to which an African defendant might plead guilty, but be unable to provide any detail of what happened, claiming to have been 'in the grip of uncontrollable passion'. Carman suggested that an explanation for these acts of violence could lie in the slower thought processes of Africans:

> Again, in murder trials when the defence is that a man killed another accidentally, intending to hurt him or to restrain him only, he may use a lethal weapon when a relatively non-lethal one is ready to hand. Such a choice is interpreted as evidence that he meant to kill and in the case of the European the inference is almost certainly justified. In the case of a Native, my personal view is that an assumption of this kind is open to discussion. It is a question of the reaction of a brain which we have no right with our present knowledge to assume to work along lines which we are taught to regard as reasonable.[109]

One implication of the notion of lower African intelligence then, was lower African criminal responsibility. Carman, in accordance with this line of thought, questioned the morality of capital punishment:

> In England, children under the age of 18 cannot be hanged and I have often wondered why if an unsophisticated African has a mentality comparable to that of a child of his own race and vastly inferior to that of an English school-boy, he should pay the ultimate penalty.[110]

The daunting ramifications of adjusting legal processes to presumed relative racial, and even tribal, capacities are demonstrated in a case in 1939, in which Carothers (who by this time had taken over responsibility for Mathari Mental Hospital) was asked to examine an African defendant, named Chemweno. Carothers claimed that Chemweno was feeble-minded, but Carman also examined him and concluded that he was 'of average intelligence for a man of his tribe, viz., Elgeyo'. Paterson discussed the matter further with Carman, who stated that if the prisoner had been a Kikuyu or a Maasai, he would have diagnosed him as feeble-minded because these tribes had a higher level of intelligence than the Elgeyo. According to Carman, the overall level of intelligence of the Elgeyo meant that, in this tribal context, Chemweno was normal, despite being mentally deficient. Paterson concluded by stating that both Carman and Carothers agreed that, for whatever reasons, Chemweno was feeble-minded and accordingly recommended mercy.[111]

The extreme sensitivity of European settlers in Kenya to the question of African criminality meant that the tendency of the theories of 'native backwardness' to encourage mercy was never going to gain ground. What is more, the complexities of such a relativistic approach, particularly when it encompassed tribal differences, made the implementation of such ideas unworkable. Carman's argument for leniency on the basis of African mental capacity reveals a theme that constantly recurred in the eugenic language of Medical Officers and other officials like La Fontaine. Carman, who described lecturing in eugenics as one of the many tasks of the Medical Officer,[112] treated the theories on race and intelligence not as a means of defending settler interests, but of protecting Africans. A combination of racist paternalism and a desire to apply modern psychological thought to social policy strongly figured in the attitudes of Kenya's eugenicist officials.

This attempt to reconcile extreme racism with social progressivism was also present in Gordon's work at Mathari Mental Hospital. Before Gordon arrived, there had been no consistent psychiatric care at Mathari, and the institution served as a place of detention rather than diagnosis and treatment, which was indicated by the fact that the statement on Mathari was placed under the section on prisons in the annual medical reports. Complaints were made in a 1932 Report of a Board of Enquiry at

Mathari about Gordon's neglect of administrative duties, the number of hours he spent at the Hospital, and his accessibility when he was urgently required,[113] although the 1935 Annual Medical Report stressed the progress that had been made at Mathari.[114] Gordon was determined to change attitudes towards mental health in Kenya; he was responsible for renaming Mathari a mental hospital rather than an asylum, a term Gordon considered outdated and as denoting only a punitive, detaining role.[115] In his attitude to the inmates at Mathari Mental Hospital Gordon displayed a serious concern about issues like the after-care of released inmates.[116] He was actively interested in the welfare of patients and the problem of over-crowding, particularly the need for more space in the African section of the hospital.[117] Gordon's work at Mathari reveals an aspect of his approach to race and social problems that is relevant to the eugenics movement in Kenya. Gordon's theories about biological racial inferiority were not considered by him to be incompatible with progressive ideas on social policy.

As Visiting Physician at Mathari, Gordon was the psychiatric expert in Kenya, a position that involved legal as well medical responsibilities. Gordon, along with Dr Violet Clarke, was a medical representative on the Special Committee formed in 1931 to consider a new Mental Disorders Bill, which it was hoped would replace the 1858 Indian Lunacy Act. An attempt had been made to introduce a new Mental Disorders Bill in 1930, but it failed because of the opposition to it generated by Gordon and the Kenyan BMA, who wanted the new legislation to be based on England's recent Mental Treatment Act of 1930, which introduced a greater element of psychiatric treatment into the legal process. The Special Committee attempted to win the co-operation of Gordon and the BMA and hasten the passage of new mental health legislation. Gordon was damning of current legislation in Kenya, the Indian Lunacy Act:

> No person in Kenya is quite as disordered as Kenya's Lunacy Act ... It is senile, lost to its surroundings, and dirty in its sections. The dirtiest thing about it is that it debars civilian doctors from certifying, throws the duty on Medical Officers, expects them to be mental experts.[118]

The existing Lunacy Law placed responsibility for the certification of insanity on the local Medical Officer and Magistrate; it was designed to enable local officials to deal with the immediate administrative and public order difficulties presented by insanity. Gordon argued that this was not enough, and that certification should be undertaken with a view to treatment and care, not simply maintaining order.[119] The old Lunacy Act, with the power it gave to local Magistrates and Medical Officers to act was suited to the colonial environment where there was a lack of psychiatric expertise and facilities. Gordon wanted the law to be changed in accordance with recent British legislation so that psychiatric intervention and treatment was central to the process of certifying and detaining of the insane. The level of reform demanded by Gordon was not easily compatible with the colonial state; it was in keeping with the levels of expertise and infrastructure available in the metropole.

Gilks, when he was still Director of Medical Services, was in agreement with Gordon. The bullishness of their stance is recorded in a conversation held with Gilks reported by the Acting Attorney-General in 1932: 'He informed me that in his opinion practically every clause of the bill would be the subject of hostile criticism either by himself or by Dr Gordon.'[120] Gordon had drawn up amendments to the bill that would make it more in line with the British legislation. He wanted to replace detention with the principle of treatment and preventive treatment and replace terms such as 'lunatic' and 'mental disorder' with 'patient' and 'mental illness'. He also demanded that the diagnosis and classification of the patient should not be based on the duration of the illness, but through an examination of the patient, securing a more definite and uniform classification of the disorders. In other words, he wanted the management of the insane to be based on modern British psychiatric practice.[121]

Gordon was uncompromising and his difficult behaviour on this issue eventually seems to have finished his career. In October 1936 the Attorney-General recommended that a 'less obstructive expert' should take the position of Visiting Physician at Mathari and that a new committee should be formed to consider a new mental disorders ordinance without Gordon on it.[122] Gordon was furious when his contract at Mathari was not renewed in 1937; he mobilised support from the Kenyan BMA, which passed a resolution urging that Gordon be retained and condemning as 'a grave discourtesy' the brusque manner of his dismissal.[123] The decision was not reversed, despite protests to the Kenya Government and the Colonial Office in London.[124] The introduction of new mental health legislation was further delayed by the outbreak of war, and it was not until 1949 that the Kenya Mental Treatment Act was passed.[125]

The main point for the purpose of this study is the link between Gordon's approach to the social policy of mental illness and eugenics. Gordon's interest in reforming and modernising the treatment of mental illness in Kenya seems at one level in conflict with the attitude to Africans conveyed by his reactionary thinking on race and intelligence. Although Dr Gordon's ideas on race fell in with most extreme settler racism, they were also motivated by a sense of the need for the most modern scientific approach to social problems—the implementation of the most up-to-date mental health policy fitted with the eugenic agenda.

[173]

Thus the extremity of Gordon's thinking was partly justified by the ideology of progress and dictated by a distorted vision of modern scientific objectivity. Eugenics was essentially the application of biological solutions to social problems; Gordon's attitude to mental health and to brain and intelligence both complied with this modern approach. The difficulty that Kenyan eugenicists found was that the limited administrative and welfare infrastructure of the colonial state did not allow the implementation of the ambitious policies envisaged for 'scientific colonization'.

Notes

1 J. Lewis, *Empire State-Building: War and Welfare in Colonial Kenya* (Oxford, 2000).
2 Gilks to the Colonial Secretary, 19 October 1932, KNA, AG/52/169.
3 Oldham to Ormsby-Gore, 11 January 1929, SOAS, IMC/CBMS/219.
4 S. J. Gould, *The Mismeasure of Man* (London, 1997), p. 205.
5 Ibid., pp. 204–22.
6 G. H. Thomson, *The Factoral Analysis of Human Ability* (London, 1939).
7 G. H. Thomson, *The Education of an Englishman: An Autobiography* (Edinburgh, 1969).
8 G. H. Thomson, *The Trend of National Intelligence: The Galton Lecture, 1946* (London, 1947).
9 'Memorandum on the proposed visit of R. A. C. Oliver to study educational methods in Africa', SOAS, IMC/CBMS/219.
10 R. A. C. Oliver, *Research in Education* (London, 1946), pp. 17–18, 51.
11 For more on the history of the Jeanes School movement, see K. J. King, *Pan-Africanism and Education: A Study of Race Philanthropy and Education in the Southern States of America and East Africa* (Oxford, 1971)
12 Lester to Ormsby-Gore, 1/10/27, NA, CO 533/372/5.
13 Memorandum on the Jeanes School by Oldham, 1926, NA, CO 533/673.
14 Education Department Annual Report, 1931, p. 32.
15 Education Department Annual Report, 1932, p. 39.
16 Education Department Annual Report, 1929, p. 49.
17 Education Department Annual Report, 1931, p. 71.
18 Transcript of a conversation with Hugh and Kathleen Trowell in 1982, p. 11, RH, MSS.Afr.s.1872/XXXIV.
19 'The Jeanes School', *EAS*, 25 July 1931, p. 11.
20 Minutes of Evidence taken before the Joint Select Committee on East Africa, 1931, NA, CO 533/408/17.
21 J. W. C. Dougall, 'Characteristics of African Thought', *Africa*, 5/3 (1932), pp. 249–65, 249. The success of this article led the IIALC to publish an extended version in pamphlet form under the same title, as IIALC Memorandum no. 10 (London, 1932).
22 Dougall, 'Characteristics of African Thought'.
23 R. A. C. Oliver, 'The Comparison of the Abilities of Races: With Special Reference to East Africa', *EAMJ*, Part I, 9/6 (1932), pp. 160–78; Part II, 9/7 (1932), pp. 193–204.
24 Ibid., p. 81.
25 Minutes of the Advisory Committee on Native Education in Tropical Africa, 26/9/29, KNA, Education/1/170.
26 Education Department Annual Report, 1930, pp. 81–2.
27 Education Department Annual Report, 1931, p. 70.
28 'General Intelligence Test for Africans', KNA, BY/26/7.
29 Oliver, 'The Comparison of the Abilities of Races', Part I.
30 Education Department Annual Report, 1932, p. 68.

31 Education Department Annual Report, 1931, p. 70.

32 Education Department Annual Report, 1932, pp. 68–9.

33 Education Department Annual Report, 1931, p. 71.

34 Oliver, 'The Comparison of the Abilities of Races', Part I, p. 160.

35 Ibid., p. 166.

36 Ibid., p. 171.

37 F. W. Vint, 'A Preliminary Note on the Cell Content of the Prefrontal Cortex of the East African Native', *EAMJ*, 9/2 (1932), pp. 30–55.

38 Oliver, 'The Comparison of the Abilities of Races', Part II, p. 198.

39 Ibid., p. 203.

40 R. A. C. Oliver, 'Mental Tests in the Study of the African', *Africa*, 7/1 (1934), pp. 40–6, 46.

41 Ibid., p. 46. P. Hetherington, in *British Paternalism and Africa* (London, 1978), p. 176, cites more articles on this subject by Oliver, published in the journal *Overseas Education*. I have so far been unable to find these articles, but the publication details given by Hetherington are: 'The Adaptation of Intelligence Tests to Tropical Africa – I and II', *Overseas Education*, 4/4 (1933), pp. 186–91, and 5/1 (1933), pp. 8–12, and 'Comparisons of Cultural Achievement', *Overseas Education*, 5/3 (1934), pp. 107–11.

42 G. Quick, 'General Intelligence Test for Africans: Manual of Direction', *Africa*, 7/2 (1934), p. 149.

43 H. S. Scott, 'A Note on the Educable Capacity of the African', *EAMJ*, 9/4 (1932), pp. 107–8.

44 D. M. Anderson, 'Policing, Prosecution and the Law in Colonial Kenya, *c.* 1905–39' in D. M. Anderson and D. Killingray (eds), *Policing the Empire: Government, Authority and Control, 1930–1940* (Manchester, 1991), pp. 183–200.

45 This corresponds to the inter-war trend described by legal historians, Morris and Read, who identified a movement from nineteenth-century Indian to recent British law. H .F. Morris and J. S. Read, *Indirect Rule and the Search for Justice: Esseays in East African Legal History* (Oxford, 1972).

46 A. Burton has shown how concerns aroused by the growth of an urban population in colonial Africa have many parallels with nineteenth-century responses to the rise of the urban working classes in Britain: 'Wahuni (The Undesirables): African Urbanisation, Crime and Colonial Order in Dar es Salaam, 1919–1961', PhD thesis (London, 2000), pp. 26–7.

47 See Native Affairs, annual reports, 1933–36.

48 KNA, Education/1/962.

49 *Crime Committee Report*, p. 7, 14, NA, CO 533/426/17.

50 *Report of the Committee on Juvenile Crime and Kabete Reformatory*, p. 3, NA, CO 533/449/3.

51 Annual Report on Kabete Reformatory, 1929, p 2, KNA, AP/1/701.

52 Annual Report on Kabete Reformatory, 1933, KNA, AP/1/699.

53 Gordon, 'Report of a Survey of the Inmates of Kabete Reformatory', CMAC SA/EUG/C.129 (also in KNA, AP/1/701).

54 Annual Report on Kabete Reformatory, 1930, KNA, AP/1/700.

55 Letter from Murray M. Jack (Registrar, Supreme Court of Kenya) to the Colonial Secretary, 23 January 1931, KNA, AP/1/700.

56 Minutes of the Meeting of the Committee of Visitors, 3 December 1930, KNA, AP/1/700.

57 Agreement – Gordon's Employment Contract, 1 January 1934, KNA, AG/32/231.

58 Gilks to the Colonial Secretary, 3 March 1931, KNA, AP/1/699.

59 Ibid.

60 Ibid.

61 'Juvenile Welfare in the Colonies'. Draft Report of the Juvenile Delinquency Sub-Committee of the Colonial Penal Administration Committee, pp. 35–6. NA, CO 885/103.

62 *Crime Committee Report*, p. 15, NA, CO 533/426/17.

63 *Crime Committee Report*, NA, CO 533/426/17.

64 Ibid., p. 1.
65 Mary M. Shaw, 'Correspondence: The Birth Weight of Africans', *EAMJ*, 9/6 (1932), p. 180.
66 M. Michael Shaw, 'Child Welfare', *EAMJ*, 8/10 (1932), pp. 184–94
67 Grant to Huxley, 8 September 1933, RH, MSS.Afr.s.2154, Box 1, File 1.
68 Invitation card for the Society for the Study of Race Improvement, 15 May 1933, CMAC/SA/EUG/C.129.
69 'The Improvement of the Human Race', *EAS*, 25 March 1933, pp. 14–15.
70 Paterson to Acting Colonial Secretary, 1 July 1933, KNA, AG/32/231.
71 Grant to Huxley, 12 July 1933, RH, MSS.Afr.s.2154/1/1.
72 *Crime Committee Report*, pp. 5–14, NA, CO 533/426/17.
73 Ibid., p. 16.
74 Ibid., p. 5.
75 Ibid., pp. 13–14.
76 For example, see E. J. Lidbetter, 'The Social Problem Group as Illustrated by a Series of East London Pedigrees', *Eugenics Review*, 24/1 (1932), pp. 7–12.
77 C. Burt, *The Young Delinquent* (London, 1925), pp. 599–600.
78 Ibid., pp. 603–5.
79 'Report on the Applicability to Kenya of Methods Pursued in Borstal and Other Reformatory Schools in England', NA, CO 533/426/17 (also in KNA, AP/1/701).
80 Grant to Huxley, 12 July 1933, RH, MSS.Afr.s.2154/1/1.
81 'Report on the Applicability to Kenya of Methods Pursued in Borstal and Other Reformatory Schools in England', p. 29, NA, CO 533/426/17.
82 Ibid., p. 30.
83 Ibid., p. 50.
84 Ibid.
85 The Juvenile Offenders Ordinance, 1933, KNA AP/1/1699.
86 *Report of the Committee on Juvenile Crime and Kabete Reformatory*, NA, CO 533/449/3 and in KNA, AP/1/699.
87 Ibid., p. 11.
88 Annual Report on Native Affairs, 1934, pp. 141–2.
89 Annual Report on Native Affairs, 1935, pp. 151–2.
90 Annual Report on Native Affairs, 1936, p. 149.
91 'The Treatment of the African Juvenile Delinquent', by the Superintendent, Kabete Approved School, p. 4, 1935, KNA, JZ/6/5.
92 NA, CO 533/451/5.
93 KNA, Kenya Prisons/1/19 and the Prisons Department Annual Report, 1937, p. 3.
94 Wisdom (Acting Director of Education) to the Colonial Secretary, 23 April 1938, KNA, AHL/2/2.
95 Chief Inspector of Approved Schools to La Fontaine, 25 November 1937, KNA, DC/NYI/2/9/4.
96 Ibid., p. 6.
97 Ibid.
98 J. L. Willcocks, 'Juvenile Crime in Kenya', p. 1, KNA, JZ/6/5.
99 Ibid., p. 2.
100 Prisons Department Annual Report, 1934, Part II, Approved Schools, p. 21.
101 Anderson, 'Policing, Prosecution and the Law', see Table 11.3, p. 195, for figures on the increasing rate of convictions.
102 Report of the Committee on Young Persons and Children (Nairobi, 1953), p. 4, KNA, BZ/2/7.
103 G. Beuschel, 'Shutting Africans Away: Lunacy, Race and Social Order in Colonial Kenya, 1910–1963', unpublished PhD thesis (London, 2001).
104 H. L. Gordon, 'Psychiatry in Kenya Colony', *Journal of Mental Science* (January, 1934), pp. 167–70.
105 H. L. Gordon, 'An Inquiry into the Correlation of Civilization and Mental Disorder in the Kenya Native' *EAMJ*, 12/11 (1936), p. 333.

106 M. Vaughan, 'Idioms of Madness: Zomba Lunatic Asylum, Nyasaland, in the Colonial Period', *Journal of Southern African Studies*, 9/2 (1983), pp. 218–38.
107 J. A. Carman, *A Medical History of the Colony and Protectorate of Kenya: A Personal Memoir* (London, 1976), pp. 74–5.
108 Memorandum by Carman, addressed to the Director of Medical Services, 29 April 1938, KNA, JZ/2/1.
109 Ibid.
110 Memorandum by Carman addressed to the Director of Medical Services, 13 June 1939, KNA, Health/1/291.
111 Paterson to the Hon. Chief Secretary, Nairobi, 1 March 1939, KNA, Health/1/291.
112 Carman, *A Medical History of the Colony and Protectorate of Kenya*, p. 40.
113 Report of the Board of Enquiry, Mathari Mental Hospital (1932), KNA, Health/2/115.
114 Annual Medical Report, 1935, p. 10
115 Annual Medical Report, 1933, p. 51.
116 Letter from Governor Byrne, 2 February 1934, KNA, AG/53/187.
117 Memorandum by Gordon, 16 November 1936, KNA, AG/32/38.
118 H. L. Gordon, 'On Certification of Mental Disorder in Kenya', *EAMJ*, 12/12 (1936), pp. 358–65, 359.
119 Ibid.
120 Acting Attorney-General to the Colonial Secretary, 17 June 1932, KNA, AG/32/41.
121 Comments by Gordon, 1931–32, KNA, AG/32/41.
122 Attorney-General to the Colonial Secretary, 8 October 1936; Attorney-General to the Colonial Secretary, 5 July 1937, KNA, AG/32/42.
123 'A Grave Discourtesy', *East Africa*, 1 April 1937.
124 Dr Gordon's Dismissal, 4 February 1937, NA, CO 822/78/19.
125 J. C. Carothers, 'The Mental Treatment Ordinance, 1949', *EAMJ*, 26/5 (1949), pp. 128–36.

CHAPTER 7

Conclusion: the decline of the eugenics empire

This book began by proposing to examine how a British idea was transplanted to the colonial environment. Kenya has been exposed as an arena of short-lived but intense eugenic activity, where the colonial context allowed a few scientists a louder voice than would have been possible in the metropole. The reasons for this lay partly in the strength of the European consensus on race and biology and partly in the nature of the colonial population, which lacked experts on heredity who could challenge Gordon and his associates' monopoly on eugenic thought. In the case of eugenics in colonial Kenya we have a clear example of the colonial frontier's tendency to distort.[1] The emphasis on the heredity of innate characteristics and the assumption of European superiority that underlay British eugenics were made elephantine by the racial agenda that dominated Kenya colony.

While accepting the distorting effect of colonial culture, it is striking that the Kenyan eugenic theories rebounded into the metropole, without their being rejected outright as grotesque colonial caricatures: Kenyan eugenics stimulated debate in Britain and drew in major thinkers there. Ultimately the causes of the failure of Kenyan eugenics lay in the nature of the relationship between the colony and Britain. Kenya's dire finances in the 1930s meant that metropolitan backing was required to develop the ideas of the Kenyan eugenicists into a major research programme. Lack of faith in the research and a reluctance to invest government money in a project with powerful political implications meant that the ambitions of the Kenyan eugenicists were resisted by the British government. This should not surprise us; the British eugenics movement, despite a pervasive influence in Britain's elite, failed to elicit official funding or implementation beyond the limited 1913 Mental Deficiency Act. More useful in assessing the significance of Dr Gordon, in particular, is the seriousness with which many of his contemporaries regarded him. Despite the doubts that arise when considering the rigour of Gordon's research, for a time he was held to be the leader of an exciting new scientific project that could supply objective answers to the troubled questions surrounding heredity, race and intelligence.

The eugenics movement in Kenya flourished in the 1930s; led by doctors and scientists, it was supported, to different degrees and for different reasons, by successive governors, the directors of Education and Health, the acting Chief Native Commissioner, as well as district commissioners. The Dean of Nairobi Cathedral, both the Managing Director and Editor of the main daily newspaper of the colony, as well as many socially influential settlers were involved in the movement. Missionaries were the only European group engaged in public life in Kenya who appear not to have been won over by the eugenic, biological explanation for 'native backwardness'.

By the late 1930s, although there had been no radical change in settler attitudes to race and no upheaval in the policy or personnel of the colonial administration, the Kenyan eugenics movement petered out. We must assume that individuals retained their eugenic beliefs, but its potency in Kenya's lore of human biology was lost. The causes of the demise of Kenyan racial eugenics lay in the financial retrenchment of the 1930s and responses in the metropole at a time when scientific racism was being increasingly undermined on both political and intellectual grounds. Without metropolitan support, Kenyan eugenics could not be sustained as a social movement. The size and composition of the Kenyan European community was such that there were not enough individuals with the intellectual and scientific interests and authority to establish an independent, self-sufficient organisation. Kenyan eugenics was forced to look to the metropole for financial, intellectual and institutional legitimacy. The demise of Kenyan eugenics is therefore intimately linked with a changing intellectual climate in Britain. Kenyan scientific discourse rather raggedly followed the metropolitan lead when it came to the role of scientific racism and the contradictory reappraisals of the roles of race and racism in science and medicine will be discussed in this conclusion.

The movement of ideas within the imperial system has been a recurrent theme of this study; the demise of British scientific racism and its effects on eugenics in Kenya will be considered here. As has been described in earlier chapters, metropolitan concerns about the political implications as well as the correctness of the role of race in science help explain the short-term failure of the Kenyan campaign in the 1930s. At the end of 1934, Ramsay MacDonald rejected the formation of a group from the Economic Advisory Committee to consider the question of East African brains. Distrust about becoming involved in such a scheme was based on political worries and strengthened by scientific opinion; Haldane's stance on the matter was often cited within the

Colonial Office as throwing the basis of the Kenyan eugenic campaign into question.

Therefore, the reasons for the failure of Kenyan eugenics lay beyond the local medical and biological forum that had created it; but the process by which the rejection of eugenic approaches to racial difference was digested in Kenya reveals how haphazard the decline of scientific racism was. There has been a tendency to view World War Two as a watershed in the historiography of scientific racism leading to the neglect of the possible continuities and developments of biological thought on race. In the first half of the twentieth century, discussions of cultural difference, human biology and society placed heavy emphasis on the importance of innate biological racial differences. Since World War Two race has disappeared from mainstream discourse in such areas. The concept of race must be understood historically; theoretical and political contexts shape the fortunes of the idea of race. The apparent demise of racism in scientific thought in the decades following World War Two should be seen as a consequence of historical circumstances, in particular the Cold War. The seriousness with which intelligent, well-meaning individuals accepted the Kenyan theories on race and intelligence should not strike us as a bizarre, historically discrete event, but serve as a reminder of the need for vigilance and critical engagement when confronted with science that has such powerful political implications, claims of objectivity notwithstanding.

The negotiation of the roles of race and heredity in the post-war medical and psychiatric discourse of Kenya indicates some important continuities and discontinuities with pre-war science. Of particular interest to this study is the jettisoning of eugenic language, while retaining many of the assumptions and some of the biological interpretations. Between 1944 and 1946, several articles were published in the *EAMJ* that returned to the issue of race and heredity. The first, written by Vint, addressed the question: 'Why has the African not developed a civilization of his own?'[2] Vint used difference in skin colour as the starting point for elaborating more complex reasons for racial differences. The effects of the African climate and sun on Africans were then examined by Vint, in the hope that this might indicate the direction in which future research in Africa should proceed. The article went on to describe the different physiological problems arising from the African environment, working on the premise that although darker skin was an adaptation to tropical conditions, there was no reason to think it might entirely negate the effects of extreme heat. Vint proceeded to discuss the effects of various physical responses to heat, such as dehydration and loss of salt, and the effects of these factors on potassium levels, the development of histamine in the skin in response to the sun, metabolic rates and vitamin

D.[3] Vint described the unwelcome physical effects of all these factors on the body to indicate how climatic conditions affected Africans, and by implication, the nature of their cultural development. This is an interesting progression in Vint's approach to race and development. His underlying assumption that Africans were backward and uncivilised remained, but the causes for this were now linked to the physiological effects of the African climate. Environmental causes rather than innate mental incapacity were regarded as the source of the problem; eugenic thought no longer had the same influence on the biology of racial difference. An editorial on Vint's work also appeared in the *EAMJ*, whose editor at that time was Dr Jex-Blake, himself a peripheral supporter of the research into race and intelligence in the 1930s. The position of the editorial tended towards the position of innate difference:

> As things are at present it looks as if all men are not equal, if one may say so without causing offence to our socialist readers, and as if humanity at large might be divided into upper and lower classes from the social or nation-building point of view.[4]

However, uncertainty about the role of heredity was also expressed; of the role of climate in determining African civilisation, Jex-Blake had this to say:

> How far this well-known tendency may be open to correction is a question for the future to decide, no doubt after full consideration of the familiar 'Nature or Nurture' lines of argument. This matter is of great importance because, in the last resort, a civilization draws its strength only from the moral sanctions upon which it rests. As the civilization improves, so, in proportion, are increasing demands made upon the moral character of its members.[5]

In 1946 a further article on race appeared in the *EAMJ*, authored by O'Brien of the Race Relations Institute. This article was significant for its attack on scientific racism and race as an immutable, biologically defined category. O'Brien described race as a recent scientific and social *construction*, and asserted that 'race is a convenient abstraction, a "solving word", used by self-interested persons and groups in order to excuse and perpetuate inequalities in the social structure'.[6] The purpose of O'Brien's article was to assist and defend the recently formed Race Relations Committee in Kenya and promote the establishment of a Race Relations Institute. It is interesting that O'Brien chose the *EAMJ* as the forum for his long and detailed defence of the Race Relations Committee. As O'Brien himself acknowledged, he emphasised the scientific side of racism in this piece at the expense of the 'humanitarian spirit' that would also be an essential component to the decision to form such an institute.[7]

O'Brien's article earned a sceptical response from the editor of the *EAMJ*. Of O'Brien's dismissal of the 'biological conception of race' in favour of an environmental analysis of difference, an editorial commented: 'This doctrine may well be regarded by many as so revolutionary as to require a good deal more proof before it is accepted.'[8] The editorial then went on to refer to the researches by Gordon and Vint on the African brain, concluding from them that 'May it not be that the African while having a good brain has a different brain? Is it best to encourage educational and social progress along European lines or would it be better to evolve a different line of progress suited to a people with great though different intellectual possibilities?'[9] In these articles and editorials, the growing momentum of the attack on British scientific racism was being digested by the Kenyan medical profession. The problem for doctors and scientists in colonial Kenya in the immediate post-war period was that they were being forced by scientific knowledge and authority to rethink earlier certainties about the biology of race and the hereditary nature of race differences. Yet they resisted abandoning the question of African difference, which was central to the Kenyan medical problematic. The solution was to locate the causes of African difference, whether in cultural development (a new concept that insinuated African intelligence), psychological stability, or physique, in the environment. This response enabled the doctors (in their own terms) to avoid accusations of racism, while leaving them free to pursue and emphasise the same interest in race. Eugenics was jettisoned in making this change in approach; particularly in its Kenyan form, eugenics was too closely associated with pre-war scientific racism.

A good example of the post-war attitudes to race and science can be found in the writings of J. C. Carothers, who became Physician of Mathari Mental Hospital in 1938, after an interlude in which John Carman, a Medical Officer with no psychiatric experience, and James Cobb, a trained psychiatrist, had stood in. Cobb was a disastrously unstable alcoholic, who was fired after it was discovered that, among other misdemeanours, he displayed the more eccentric patients of Mathari to drinking companions on late-night tours. Carothers had previously been a Medical Officer at Kisii,[10] and had had no training in psychiatry, although later he took a diploma in Psychological Medicine at the Maudsley Hospital in London in 1946. On his retirement from the medical service in 1951, Carothers became Psychiatric Specialist at St James's Hospital in Portsmouth, and in 1952 he was commissioned to write a monograph for the World Health Organisation, *The African Mind in Health and Disease*.[11] In 1954 Carothers was appointed to write a report on Mau Mau from a psychological perspective. In his report, Carothers described Mau Mau as an expression of African psychopathy.

Carothers is a notorious figure in the history of colonial psychiatry and other historians, in particular McCulloch, have discussed his work in detail within the history of colonial psychiatry.

Carothers' ideas are of interest here for their negotiation of the relative roles of nature and nurture in post-war colonial scientific racism. Carothers' relationship with the Kenyan eugenicists and their theories was an ambiguous one. His work in African psychiatry was clearly influenced by Gordon and Vint, and their articles are often referenced in Carothers' published work. Yet Carothers placed more influence on the role of environment in shaping African psychology than innate brain structure. In his first article on the subject of psychology, printed in 1939, Carothers wrote:

> It is no doubt necessary in a given case of insanity to search for hereditary factors, infantile trauma, mental conflicts, etc., to explain the primary cause; but I suggest that the type of insanity that develops, be it schizophrenic or manic-depressive, depends on one's attitude to life. Moreover, and this is the point, this attitude is mainly dependent on one's culture.
>
> That the whole question of type of insanity is not a racial one, but a cultural one, is surely demonstrated by the remarkable variations that occur within Europe, and even within the separate countries of Europe.[12]

Carothers also went on to discuss the role of heredity in African mental illness in his 1947 article on mental derangement in Africans in the *Journal of Mental Science*, and concluded that '(rather surprisingly) ... the African's inherited tendencies to mental disease are probably not very different from the European's'.[13] Carothers explained African madness as arising from the contradictions between traditional African social structures, and the attitudes to life that such a system created, and Western social attitudes. He saw the most problematic and psychologically destabilising contrasts in the differing concepts of identity and personal responsibility. Carothers described notions of individuality, with its concomitant traits of guilt, self-sufficiency and 'a higher minimum level of intelligence'. In contrast, African social structures were described as providing detailed, external and concrete restraints on behaviour that tended to bypass the abstraction of social concepts and as a result 'internal consistency is not developed or required, and phantastic thinking prevails'.[14] Although Carothers drew an analogy between Western 'directed thinking' and evolutionary progress by citing Huxley's comment that evolutionary progress always involves greater control of the environment, he also commented that no society should allow all responsibility to be devolved to the individual, and that a reversion to 'primitive ways' in the form of 'social protection and control' would be beneficial.

The form of Carothers' scientific racism, for the purpose of analysing the role of eugenic thought in Kenyan science, was significantly different from that of the 1930s. The African mind was still treated as pathological, indeed often as psychopathological, by Carothers, but there was far heavier emphasis on the environmental effects, in the shape of social structure and physical environment, as an explanation of African psychology. The issue of nature and nurture was still addressed by Carothers, and he did not discount the role of heredity in racial differences. For example, in 1951 he published an article (again in the *Journal of Mental Science*) comparing the thinking of leucotomised Europeans with normal Africans.[15] Interestingly, this article arose from an appeal by Vint for tests to find reliable Africans to work in the Medical Research Laboratory. The tests that Carothers performed indicated to him similarities between Africans and leucotomised Europeans, which led to consideration of the 'neurophysiological basis of African thinking and character, and about the functions of the frontal lobes in general'. Carothers was returning to similar territory as the overtly eugenic research on African brains conducted in the 1930s. This article concluded:

> on the evidence of leucotomy in Europeans, that all observed African peculiarities can be explained as due to a relative idleness in his frontal lobes.
>
> This frontal idleness in turn can be accounted for on cultural grounds alone, but the possibility of anatomical differences, is not thereby excluded.[16]

Carothers' writing on African mentality was produced at a time when racial tensions were becoming increasingly serious as the Mau Mau war unfolded. Although commissioned to write on Mau Mau, Carothers never engaged with the political context – all tensions were internalised inside an inadequate and troubled African mind. Politicisation was treated as a manifestation of psychological disorder, as a monumental process of acting out by a people too immature to resolve its own internal conflicts arising from a psychological inability to deal with modern life. A similar refusal to accept the validity of African politicisation can be seen in the theories of the Kenyan eugenicists about limited African educational capacity at a time when the pursuit of independent education became a major issue in the early 1930s. Carothers' work on Mau Mau was an important and influential expression of scientific racism in Kenya; the nature of the science is interesting in its greater uncertainty about the role of racial heredity and emphasis on environmental factors.[17] Carothers was not the only European scientist in Africa adjusting the emphasis on nature and nurture in an attempt to avert accusations of racism; as Dubow has argued, theoreticians of apartheid after 1948 'adopted ambiguous attitudes towards scientific racism and ... were mostly careful to couch their explanations of racial differences in cultural and religious rather than biological terms'.[18]

Eugenics was therefore swiftly, though not entirely convincingly, displaced as the key to racial difference in post-war Kenyan medical discourse. A similar process was taking place in Britain, but in reverse, where the role of race and scientific racism in eugenics was downplayed. In this book, these more recent processes in the intellectual history of eugenics and scientific racism are scraped back to give an account of eugenics and colonial scientific racism in their time and place. In doing this, I have attempted to demonstrate how influences from Britain and the colony interacted to create a distinctive and extreme eugenic agenda in the imperial environment. In examining this relationship, Kenya held a mirror up to British eugenics, revealing how the problem of racial difference was implicitly at the heart of its hereditarian worldview.

Notes

 1 See H. Lamar and L. Thompson, *The Frontier in History: North America and Southern Africa Compared* (New Haven, 1981), especially Chapter 4, 'Processes in the Development of the Southern African Frontier' by H. Giliomee, pp. 76–119, and I. Kopytoff, *The African Frontier: The Reproduction of Traditional African Societies* (Bloomington, 1987).
 2 F. W. Vint, 'Solar Rays. Fact or Fiction?', *EAMJ*, 21/8 (1944), pp. 227–39, 227.
 3 Ibid., p. 227.
 4 Editorial, 'African Civilizations', *EAMJ*, 21/8 (1944), pp. 222–6, 225.
 5 Ibid., p. 226.
 6 O'Brien, 'The Methods and Aims of a Race Relations Institute', *EAMJ*, 23/12 (1946), pp. 361–84, 361.
 7 Ibid., p. 382.
 8 Editorial, *EAMJ*, 23/12, pp. 359–60, 359.
 9 Ibid., p. 360.
10 J. A. Carman, *A Medical History of the Colony and Protectorate of Kenya: A Personal Memoir* (London, 1976), pp. 74–5.
11 J. McCulloch, *Colonial Psychiatry and 'the African Mind'* (Cambridge, 1995), p. 50; J. C. Carothers, *The African Mind In Health and Disease. A Study in Ethnopsychiatry* (Geneva, 1953).
12 J. C. Carothers, 'Some Speculations on Insanity in Africans, and in General', *EAMJ*, 17/3 (1940), pp. 90–105, 102–3.
13 J. C. Carothers. 'A Study of Mental Derangement in Africans, and an Attempt to Explain Its Peculiarities, More Especially in Relation to the African Attitude to Life', *Journal of Mental Science*, 93 (1947), pp. 548–97, 594.
14 Ibid., p. 595.
15 J. C. Carothers, 'Frontal Lobe Function and the African', *Journal of Mental Science*, 97 (1951), pp. 12–48.
16 Ibid., pp. 46–7.
17 J. C. Carothers, *The Psychology of Mau Mau* (Nairobi, 1954).
18 S. Dubow, *Scientific Racism in Modern South Africa* (Cambridge, 1995), p. 288.

ABBREVIATIONS IN NOTES

BMA	British Medical Association
BMJ	*British Medical Journal*
CMAC	Contemporary Medical Archives Centre
CO	Colonial Office
EAS	*East African Standard*
EAMJ	*East African Medical Journal*
ES	Eugenics Society
IMC/CBMS	International Missionary Council/Conference of British Missionary Societies
KEAMJ	*Kenya and East African Medical Journal*
KNA	Kenya National Archives
KSSRI	Kenya Society for the Study of Race Improvement
NA	National Archives
RH	Rhodes House Library, Oxford

BIBLIOGRAPHY

Archives and libraries consulted

Bodleian Library, Oxford
Grigg Papers

British Medical Association Archives, London

Colindale Newspaper Library, London

Eugenics Society Archive, held at the **Contemporary Medical Archives Centre, Wellcome Institute for the History of Medicine** (please note: all material from the Eugenics Society Archive © The Galton Institute, London)

House of Lords Records Office, London

Kenya National Archives, Nairobi

The National Archives, Kew, London (formerly the Public Record Office)

Rhodes House Library, Oxford

School of Hygiene and Tropical Medicine Library, London
Kenya and East African Medical Journal/East African Medical Journal

School of Oriental and African Studies Library, London

Official publications

Royal Commission on the Care and Control of the Feeble-Minded, *Minutes of Evidence and Reports*, Cmd 4215–4221 (London: HMSO, 1908).

Advisory Committee on Native Education in the British Tropical African Dependencies, *Education Policy in British Tropical African Dependencies*, Cmd 2374 (London: HMSO,1925).

Report of the Mental Deficiency Committee being a Joint Committee of the Board of Education and the Board of Control (Wood Report) (London: HMSO, 1929).

Report of the Commission on Closer Union in Eastern and Central Africa (Hilton Young Report), Cmd 3234 (London: HMSO, 1929).

Memorandum on Native Policy, Cmd 3573 (London: HMSO, 1930).

Report of the Joint Select Committee on Closer Union in East Africa, H. C. 156 (London: HMSO, 1931).

Crime Committee Report (Nairobi: Government Printers, May 1932).

Report of the Commission of Inquiry into the Administration of Justice in Kenya, Uganda and the Territory of Tanganyika (Bushe Report), Cmd 4623 (London: HMSO, 1934).

Report of the Committee on Juvenile Crime and Kabete Reformatory (Nairobi, Government Printers, 1934).

Board of Control, Committee of Sterilisation, *Report (Brock Report)*, Cmd 4485 (London: HMSO, 1934).

L. S. Penrose, *A Clinical and Genetic Study of 1280 Cases of Mental Defect (Colchester Survey)*, No. 229 in Special Report series (London: HMSO, 1938).

H. Slade et al., *Report of the Committee on Young Persons and Children* (Nairobi: Government Printers, 1953).

Primary published literature

Aikman, K. B., 'Race Mixture', *Eugenics Review*, 25/3 (1933): 161–66.

Arnell, O. R., 'Correspondence', *East African Medical Journal*, 9/6 (1932): 179–80.

Bateson, W., 'Commonsense in Racial Problems', *Eugenics Review*, 13/1 (1921): 325–38.

Berry, R. J. A, 'Practical Method for the Detection, during Childhood, of Potential Social Inefficiency and High-Grade Mental Deficiency' in *Stoke Park Monographs on Mental Deficiency and Other Problems of the Human Brain and Mind*, edited by Berry, R. J. A., pp. 3–39, London: Macmillan, 1933.

Berry, R. J. A, 'Mental Deficiency in England: An Analysis of the Mental, Physical and Mental Characteristics of a Group of 162 Adult Feeble-Minded Women' in *Stoke Park Monographs on Mental Deficiency and Other Problems of the Human Brain and Mind*, edited by Berry, R. J. A., pp. 65–80, London: Macmillan, 1933.

Berry, R. J. A., 'Unselected Examples of the Hereditary Transmission of Endogenous Amentia' in *Stoke Park Monographs on Mental Deficiency and Other Problems of the Human Brain and Mind*, edited by Berry, R. J. A., pp. 213–19, London: Macmillan, 1933.

Berry, R. J. A (ed.), *Stoke Park Monographs on Mental Deficiency and Other Problems of the Human Brain and Mind*, London: Macmillan, 1933.

Burt, Cyril, *The Young Delinquent*, London: University of London Press, 1925.

Campbell, C., 'Juvenile Delinquency in Colonial Kenya, 1900–1939', *The Historical Journal*, 45/I (2002): 129–51.

Carman, J. A. and Roberts, M. A. W., 'Social and Health Conditions among the Jaluo, with Special Reference to Maternal and Infant Welfare', *East African Medical Journal*, 11/4 (1934): 107–24.

Carothers, J. C., 'A Comparison of Two Methods of Skin Grafting', *East African Medical Journal*, 13/11 (1937): 345.

Carothers, J. C., 'The Female Kikuyu Pelvis in Relation to Labour', *East African Medical Journal*, 14/8 (1937): 260–1.

Carothers, J. C., 'Some Speculations on Insanity in Africans, and in General', *East African Medical Journal*, 17/3 (1940): 90–105.

Carothers, J. C., 'Miraa as a Cause of Insanity', *East African Medical Journal*, 22/1 (1945): 4–6.

Carothers, J. C., 'A Study of Mental Derangement in Africans, and an Attempt to Explain Its Peculiarities, More Especially in Relation to the African Attitude to Life', *Journal of Mental Science*, 93 (1947): 548–97.

Carothers, J. C., 'The Mental Treatment Ordinance, 1949', *East African Medical Journal*, 26/5 (1949): 128–36.

Carothers, J. C., 'Frontal Lobe Function and the African', *The Journal of Mental Science*, 97 (1951): 12–48.

Carothers, J. C., *The African Mind in Health and Disease: A Study in Ethnopsychiatry*, Geneva: World Health Organization, 1953.

Carothers, J. C., *The Psychology of Mau Mau*, Nairobi: Government Printers, 1954.

Carothers, J. C., *The Mind of Man in Africa*, London: Tom Stacey, 1972.

Cook, A., 'Note on Dr. Mitchell's Paper on the Causes of Obstructed Labour in Uganda', *East African Medical Journal*, 15/7 (1938): 213–17.

Cook, W. G. H., 'English and Foreign Marriage Law in Relation to Mental Disorder', *Eugenics Review*, 13/1 (1921): 351–9.

Crewdson Benington, R., 'A Study of the Negro Skull with Special Reference to the Congo and Gaboon Crania', *Biometrika*, 8/3 and 4 (1912): 292–337.

Crookshank, F. G., *The Mongol in Our Midst: A Study of Man and His Three Races*, London: Kegan Paul, 1931.

Dahlberg, G., *Race, Reason and Rubbish*, translated by L. Hogben, London: George Allen and Unwin, 1942.

Dougall, J. W. C., 'Characteristics of African Thought', *Africa*, 5/3 (1932): 249–65.

Dougall, J. W. C., 'The Development of the Education of the African in Relation to Western Contact', *Africa*, 11/3 (1938): 312–23.

Dover, C., 'Population Control in India', *Eugenics Review*, 26/4 (1934): 283–5.

Duerden, J. E., 'Genesis and Reclamation of the Poor White in South Africa', *Eugenics Review*, 14/4 (1923): 270–5.

EAMJ, 'Kenya Society for the Study of Race Improvement', *East African Medical Journal*, 10/4 (1933): 155–6.

EAMJ, 'Kenya Society for the Study of Race Improvement', *East African Medical Journal*, 10/12 (1934): 379.

EAMJ, 'The Kampala Meeting: Resolution on Research in Africa', *East African Medical Journal*, 13/4 (1936): 122.

EAMJ, 'Mental Degeneracy', *East African Medical Journal*, 16/4 (1939): 190–1.

EAMJ, 'East African Medical Service Lists', *East African Medical Journal*, 16/8 (1939): 312–19.

Editorial, *British Medical Journal*, 30 April (1932): 812–13.

Editorial, 'The Mind and Brain of the Kenya Native', *British Medical Journal*, 18 November (1933): 931–2.

Editorial, 'The Medical Aspect of Closer Settlement of Europeans in the Kenya Highlands' *Kenya and East African Medical Journal*, 6/7 (1929): 188–9.

Editorial, *East African Medical Journal*, 9/2 (1932): 29.

Editorial, *East African Medical Journal*, 9/10 (1933): 275.

Editorial, *East African Medical Journal*, 10/12 (1934): 349.

Editorial, 'The Brain of the East African', *East African Medical Journal* 11/2 (1934): 39–42.

Editorial, *East African Medical Journal*, 11/7 (1934): 209–10.

Editorial, 'The Medical Education of Africans', *East African Medical Journal*, 16/5 (1939): 167.

Editorial, *East African Medical Journal*, 19/1 (1942): 1–2.

Editorial, 'African Civilizations', *East African Medical Journal*, 21/8 (1944): 225.

Editorial, 'The Training of Native Laboratory Assistants in East Africa', *East African Medical Journal*, 21/3 (1944): 65–6.

Editorial, *East African Medical Journal*, 23/12 (1946): 359–60.

Editorial, 'The Brain of the African Native', *Lancet*, 226/5764 (1934): 357.

Fawcett, C. D., assisted by A. Lee and Others, 'A Second Study of the Variation and Correlation of the Human Skull, with Special Reference to the Naqada Crania', *Biometrika*, 1/4 (1902): 408–67.

Fick, M. L., *The Educability of the South African Native, Issued by the South African Council for Educational and Social Research*, Pretoria: The National Bureau, 1939.

Fleming, R. M., 'Human Hybrids', *Eugenics Review*, 21/4 (1930): 257–63.

Fleure, H. J., 'Some Aspects of Race Study', *Eugenics Review*, 14/2 (1922): 93–102.

Fleure, H. J., 'The Nordic Myth', *Eugenics Review*, 22/2 (1930): 117–21.

Fleure, H. J., 'Race and Politics', *Eugenics Review*, 27/4 (1936): 319–20.

Foiley, E. J., 'Psychiatric Opinion and Crime with Special Reference to Temporary Insanity', *East African Medical Journal*, 23/5 (1946): 150–4.

Galton, F., *Hereditary Genius*, London: Macmillan, 1869.

Galton, F., *Essays in Eugenics*, London: Eugenics Education Society, 1909.

Gates, R. R., 'Heredity and Eugenics. Part II', *Eugenics Review*, 12/1 (1920): 1–13.

Gates, R. R., 'Eugenics and Education', *Eugenics Review*, 23/4 (1932): 305–9.

Gibbons, R. M. and M. J., 'Some Considerations for the Training of African Medical Students', *East African Medical Journal*, 10/11 (1934): 355–7.

Gilks, J. L., 'The Medical Department and the Health Organization in Kenya, 1909–1923', *East African Medical Journal*, 9/12 (1933): 340–54.

Gilks, J. L., 'Dietetic Problems in East Africa', *East African Medical Journal*, 10/9 (1933): 254–65.

Gilks, J. L., 'The Relation of Economic Development to Public Health in Rural Africa', *Journal of the African Society*, 34/134 (1935): 31–40.

Gilks, J. L. and J. B. Orr, 'The Nutritional Condition of the East African Native', *Kenya and East African Medical Journal*, 4/3 (1927): 85–90.

Gordon, H. L., *Sir James Simpson and Chloroform*, London: T. Fisher & Unwin, 1897.

Gordon, H. L., *The Modern Mother: A Guide to Girlhood, Motherhood and Infancy*, London: T. Werner Laurie, 1909?

Gordon, H. L., *The Prevention of Malaria in the Northern Transvaal*, Johannesburg: Argus, 1910.

Gordon, H. L., 'Mental Instability among Europeans in Kenya', *Kenya and East African Medical Journal*, 4/10 (1928): 316–24.

Gordon, H. L., 'Paranoia; and the Case of Hackman', *Kenya and East African Medical Journal*, 5/11 (1929): 874–6.

Gordon, H. L., '(1) A Case of Spatial Syndrome (2) Sexual Perversions', *Kenya and East African Medical Journal*, 6/5 (1929): 122–30.

Gordon, H. L., 'Relation of Malaria to the Alleged Rarity of Neurosyphilis amongst "Uncivilised" Races', *Kenya and East African Medical Journal*, 6/8 (1929): 221–9.

Gordon, H. L., 'Some Implications of the Closer Union Report', *East African Standard*, 13 April (1929): 15–16.

Gordon, H. L., 'Notes on Some Current Literature and Topics', *Kenya and East African Medical Journal*, 7/4 (1930): 108–16.

Gordon, H. L., 'A Note on the Diagnosis of Amentia (Mental Deficiency) in Africans', *Kenya and East African Medical Journal*, 7/8 (1930): 208–14.

Gordon, H. L., 'The Social Aspect of the Curious History of Syphilis', *Kenya and East African Medical Journal*, 7/11 (1931): 320–2.

Gordon, H. L., 'Correspondence. The Educable Capacity of the African', *East African Medical Journal*, 9/7 (1932): 210–12.

Gordon, H. L., 'Xeropthalmia in Mathari Mental Hospital', *East African Medical Journal*, 10/3 (1933): 85–90.

Gordon, H. L., 'The Case of Charles William Ross, Hanged for Murder in Kenya Colony' in *Stoke Park Monographs on Mental Deficiency and Other Problems of the Human Brain and Mind*, edited by Berry, R. J. A., pp. 233–7, London: Macmillan, 1933.

Gordon, H. L., 'Amentia in the East African', *Eugenics Review*, 25/4 (1934): 223–35.

Gordon, H. L., 'The Basal Metabolism of the East African (Letter to the Editor)', *East African Medical Journal*, 11/4 (1934): 134.

Gordon, H. L., 'The Intentional Improvement of Backward Tribes', *East African Medical Journal*, 11/5 (1934): 143–56.

Gordon, H. L., 'The Kenya Native', *Eugenics Review*, 26/1 (1934): 85–7.

Gordon, H. L., 'Brain and Mind', *Eugenics Review*, 26/4 (1934): 311.

Gordon, H. L., 'Neurospirochætosis in the East African', *Proceedings of the Royal Society of Medicine*, 27/Part 1 (1934): 243–54.

Gordon, H. L., 'The Mental Capacity of the African', *Journal of the African Society*, 33/132 (1934): 226–42.

Gordon, H. L., 'Psychiatry in Kenya Colony', *Journal of Mental Science* (January, 1934): 167–70.

Gordon, H. L., 'Yaws and Syphilis, Correspondence', *East African Medical Journal*, 12/7 (1935): 221–2.

Gordon, H. L., 'An Inquiry into the Correlation of Civilization and Mental Disorder in the Kenya Native', *East African Medical Journal*, 12/11 (1936): 327–35.

Gordon, H. L., 'On Certification of Mental Disorder in Kenya', *East African Medical Journal*, 12/12 (1936): 358–65.

Gordon, H. L., 'Headache in Kenya', *East African Medical Journal*, 14/6 (1937): 189–98.

Gordon, H. L., 'The Female Baganda Pelvis', *East African Medical Journal*, 15/8 (1938): 268–69.

Gordon, H. L., 'The Doctor in the Witness-Box', *East African Medical Journal*, 18/8 (1941): 236–39.

Gordon, H. L., 'Is War Eugenic or Dysgenic? That is Does War Improve or Impair the Physical or Mental Qualities of Future Generations', *East African Medical Journal*, 19/2 (1942): 86–96.

Gordon, H. L., 'The Importance of Social Medicine to Kenya', *East African Medical Journal*, 23/1 (1946): 2–12.

Graham-Little, E., 'British Empire and Backward Races', *Empire Review and Magazine*, 66/442 (1937): 269–75.

Gregory, J. R., 'Ante–Natal Care and Carelessness', *East African Medical Journal*, 16/3 (1939): 104–25.

Grigg, E., *The Constitutional Problem in Kenya*, Nottingham: University College, Nottingham, 1933.

Hailey, M., *An African Survey*, London: Oxford University Press, 1938.

Haldane, J. B. S., *The Causes of Evolution*, London: Longman, Green and Co., 1932.

Haldane, J. B. S., *The Inequality of Man and Other Essays*, London: Chatto and Windus, 1932.

Haldane, J. B. S., *Fact and Faith*, London: Watts and Co., 1934.

Haldane, J. B. S., *Human Biology and Politics*, London: The British Science Guild, 1934.

Haldane, J. B. S., *Heredity and Politics*, London: George Allen and Unwin, 1938.

Hankins, F. H., 'Civilization and Fertility. Has the Reproductive Power of Western Peoples Declined?', *Eugenics Review*, 23/2 (1931): 145–50.

Harvey, D. and Vint, F. W., 'A Note on the Post-Mortem Calcium Content of the Blood Serum and Cerebro-Spinal Fluid of the East African Native', *Kenya and East African Medical Journal*, 8/9 (1931): 240–5.

Healy, W., *The Individual Delinquent*, New York: Alfred A. Knopf, 1914.

Hentachri, C. C., 'South African Journal of Science', *Eugenics Review*, 21/1 (1929): 78.

Hentschel, C. C., 'South African Journal of Science', *Eugenics Review*, 23/1 (1931): 91–92.

Hinde, H., 'The "Black Peril" in British East Africa: A Frank Talk to Women Settlers', *The Empire Review and Journal of British Trade*, 35/245 (1921): 193–200.

Hogben, L., *Genetic Principles in Medicine and Social Science*, London: Williams and Norgate, 1931.

Hogben, L., *Nature and Nurture*, London: Williams and Northgate, 1933.

Hogben, L., *The Retreat from Reason*, London: Watts and Co., 1936.

Holmes, S. J., 'Natural Selection in Man', *Eugenics Review*, 22/1 (1930): 7–16.

Hubback, E. and Green, M. E., 'Family Endowment', *Eugenics Review*, 25/1 (1933): 33–6.

Huxley, J. and Haddon, A. C., *We Europeans: A Survey of 'Racial Problems'*, London: Jonathan Cape, 1935.

Huxley, J., *Africa View*, London: Chatto & Windus, 1931.

Huxley, J., 'Eugenics and Society', *Eugenics Review*, 28/1 (1936): 11–31.

Huxley, J., *Memories*, Vol. I, London: Allen and Unwin, 1970.

Hyam, R., *Empire and Sexuality: The British Experience*, Manchester: Manchester University Press, 1992.

Josey, C. C., *Race and National Solidarity*, New York: Charles Scribner and Sons, 1923.

Kenyatta, J., *Facing Mount Kenya*, London: Secker and Warburg, 1938.

Langdon Brown, W., 'The Mongol in Our Midst', *Eugenics Review*, 23/1 (1931): 251–3.

Latham, D. V., 'The White Man in East Africa', *East African Medical Journal*, 9/10 (1933): 276–82.

Leakey, L. S. B., *Mau Mau and the Kikuyu*, London: Methuen, 1952.

Leakey, L. S. B., *Defeating Mau Mau*, London: Methuen, 1955.

Leys, N., *Kenya*, London: The Hogarth Press, 1924.

Lidbetter, E. J., 'The Social Problem Group as Illustrated by a Series of East London Pedigrees', *Eugenics Review*, 24/1 (1932): 7–12.

Loewenthal, L. J. A., 'The Significance of Colour Changes in African Skin', *East African Medical Journal*, 11/4 (1934): 124–31.

Loewenthal, L. J. A., 'The Effect of the Addition of Milk to the Diet of Schoolboys in Buganda', *East African Medical Journal*, 15/1 (1938): 35–45.

Ludovici, A. M., 'Feminism and Science', *Eugenics Review*, 21/1 (1929): 50–1.

Lundborg, H., 'The Danger of Degeneracy', *Eugenics Review*, 13/4 (1921): 531–9.

MacCrone, I. D., *Race Attitudes in South Africa*, London: Oxford University Press, 1937.

Mackay, D., 'A Background for African Psychiatry', *East African Medical Journal*, 25/1 (1948): 2–4.

Mackinnon, M., 'Medical Aspects of White Settlement in Kenya', *East African Medical Journal*, 11/12 (1935): 376–93.

Mayhew, A., 'A Comparative Survey of Educational Aims and Methods in British India and British Tropical Africa', *Africa*, 6/2 (1933): 172–86.

McCardie, Justice, 'My Outlook on Eugenics', *Eugenics Review*, 25/1 (1933): 7–14.

McDougall, W., *National Welfare and National Decay*, London: Methuen, 1921.

Michener, R. B., 'Correspondence. The Birth Weight of Africans', *East African Medical Journal*, 9/5 (1932): 149.

Michener, R. B., 'Queries on Native Circumcision', *East African Medical Journal*, 13/11 (1937): 378–81.

Mitchell, J. P., 'On the Causes of Obstructed Labour in Uganda', *East African Medical Journal*, 15/6 (1938): 177–89.

Mitchell, J. P., 'On the Causes of Obstructed Labour in Uganda. Part II', *East African Medical Journal*, 15/7 (1938): 206–12.

Mjoen, J. A., 'Race-Crossing and Glands. Some Human Hybrids and their Parental Stock', *Eugenics Review*, 23/1 (1931): 31–40.

Mockerie, P., *An African Speaks for His People*, London: The Hogarth Press, 1934.

Moore, E., 'Amentia in South Africa', *Eugenics Review*, 23/3 (1931): 239.

Muwazi, E. and Trowell, H., 'Neurological Disease among African Natives of Uganda: A Review of 269 Cases', *East African Medical Journal*, 21/1 (1944): 2–19.

Nature, 'European Civilisation and African Brains', *Nature*, 132/3347 (1933): 958.

Nielsen, P., *The Black Man's Place in South Africa*, Cape Town: Juta and Co., 1922.

Notes and Memoranda, 'The African Brain', *Eugenics Review*, 26/1 (1934): 51.

O'Brien, T. P., 'The Methods and Aims of the Race Relations Institute', *East African Medical Journal*, 23/12 (1946): 361–84.

Oldham, J. H., 'The Educational Work of Missionary Societies', *Africa*, 7/1 (1934): 47–59.

Oliver, R. A. C., 'The Comparison of the Abilities of Races: With Special Reference to East Africa', *East African Medical Journal*, 9/6 (1932): 160–75.

Oliver, R. A. C., 'The Comparison of the Abilities of Races: With Special Reference to East Africa [Concluded]', *East African Medical Journal*, 9/7 (1932): 193–204.

Oliver, R. A. C., 'Mental Tests in the Study of the African', *Africa*, 7/1 (1934): 40–46.

Oliver, R. A. C., *The Training of Teachers in Universities*, Bickley (wartime address): University of London Press, 1943.

Oliver, R. A. C., *Research in Education*, London: Allen and Unwin, 1946.

Onslow, H., 'Fair and Dark; Is There a Predominant Type?', *Eugenics Review*, 12/3 (1920): 212–17.

Orr, J. B., *Food, Health and Income. Report on the Inadequacy of Diet in Relation to Income*, London: Macmillan, 1936.

Paterson, A., 'Nutrition and Agriculture', *East African Medical Journal*, 17/2 (1940): 51–9.

Paterson, A. R., 'The Education of Backward Peoples', *East African Medical Journal*, 8/11 (1932): 302–15.

Paterson, A. R., *On Various Public Health Questions*, Health Pamphlet No. 13: Colony and Protectorate of Kenya, Medical Department, 1937.

Pearl, R., 'Variation and Correlation in Brain Weight', *Biometrika*, 4/1 and 2 (1905): 13–104.

Pearson, K., *National Life from the Standpoint of Science*, London: Black and Co., 1901.

Pearson, K., *Social Problems: Their Treatment, Past, Present and Future*, London: Cambridge University Press, 1912.

Pearson, K., *Eugenics and Public Health. An Address to Public Health Officers*, London: Dulau and Co., 1912.

Pearson, K., *The Function of Science in the Modern State*, Cambridge: Cambridge University Press, 1919.

Penrose, L. S., *Mental Defect*, London: Sidgwick and Jackson, 1933.

Penrose, L. S., 'The Complex Determinants of Amentia', *Eugenics Review*, 26/2 (1934): 121–6.

Perham, M. (ed.), *Ten Africans*, London: Faber and Faber, 1936.

Philip, C. R., 'Nutrition in Kenya: Notes on the State of Nutrition of African Children', *East African Medical Journal*, 20/7 (1943): 227–34.

Pitt-Rivers, G. H. L.-F., 'Sex-Ratios and Marriage', *Eugenics Review*, 21/1 (1929): 21–8.

Pocock, H. A., 'Sterilization in the Empire. An Account of the Working of the Alberta Act', *Eugenics Review*, 24/2 (1932): 127–30.

Preston, P. G., 'Notes on Two Years' Maternity Work amongst the South Kavirondo (Kenya) Natives (215 cases)', *East African Medical Journal*, 13/7 (1936): 215–23.

Quick, Griffith, 'General Intelligence Test for Africans: Manual for Direction', *Africa*, 7/2 (1934): 149.

Raymond, W., 'Minimum Dietary Standards for East African Natives', *East African Medical Journal*, 17/7 (1940): 249–65.

Ritchie, J. F., *The African as Suckling and as Adult (A Psychological Study)*, The Rhodes-Livingstone Papers, Number 9, The Rhodes-Livingstone Institute, 1943.

Roberts, J. and Dick, D. A., 'Notes on the Control of Bed-Bugs (*Cimex Rotundatus*)', *East African Medical Journal*, 12/1 (1935): 46–57.

Saleeby, C. W., *Parenthood and Race Culture*, London: Cassell, 1909.

Schiller, F. C. S., 'Eugenics versus Civilization', *Eugenics Review*, 13/2 (1921): 381–93.

Schofield, A. T., 'Scientific Diets for African Children', *East African Medical Journal*, 13/8 (1936): 230–45.

Schuster, E. and Elderton, E., *The Inheritance of Ability, being a Statistical Study of the Oxford Class Lists and of the School Lists of Harrow and Charterhouse*, London: Dulau and Co., 1907.

Scott, H. S., 'A Note on the Educable Capacity of the African', *East African Medical Journal*, 9/4 (1932): 99–110.

Sequeira, J. H., 'The Brain of the East African Native', *British Medical Journal*, 1 (1932): 581.

Sequeira, J. H., 'The Influence of Light and Heat on the Human Body', *East African Medical Journal*, 8/12 (1932): 332–58.

Sequeira, J. H., 'Correspondence. The Educable Capacity of the African, *East African Medical Journal*, 9/5 (1932): 146–8.

Sequeira, J. H., 'Heredity and Disease: The Story of the Abbé Gillet's Rabbits', *East African Medical Journal*, 10/11 (1934): 331–5.

Shaw Bolton, J., *The Brain in Health and Disease*, London: Edward Arnold, 1914.

Shaw Bolton, J., *The Cortical Localisation of the Cerebral Function*, Edinburgh: Oliver and Boyd, 1933.

Shaw, M. Michael, 'Child Welfare', *East African Medical Journal*, 8/10 (1932): 284–94.

Shaw, M. Michael, 'The Birth-Weight of Africans', *East African Medical Journal*, 10/1 (1933): 32–4.

Shaw, Mary M., 'Correspondence. The Birth Weight of Africans', *East African Medical Journal*, 9/6 (1932): 180.

Sherrington, C., *The Brain and Its Mechanism*, Cambridge: Cambridge University Press, 1933.

Smartt, C. G. F., 'Mental Maladjustment in the East African', *Journal of Mental Science*, 102 (1956): 441–66.

Speller, C., *Great Britain's Reward to the Kenya Native Ex-Service Man*, Aberystwyth, 1931.

Stock, C. S., 'The Sex Ratio and Emigration', *Eugenics Review*, 10/3 (1918): 163–6.

Stones, R. Y., 'Twenty-Five Years of Tropical Medicine', *East African Medical Journal*, 13/9 (1936): 271–81.

Stones, R. Y., 'On the Causes of Obstructed Labour in Uganda', *East African Medical Journal*, 15/7 (1938): 217–19.

Terman, Lewis, *The Intelligence of School Children: How Children Differ in Ability, the Use of Mental Tests in School Grading and the Proper Education of Exceptional Children*, London: Harrap and Co., 1921.

Thomson, G. H., *A Modern Philosophy of Education*, London: George Allen and Unwin, 1929.

Thomson, G. H., *Instinct, Intelligence, and Character: An Educational Psychology*, London: Allen and Unwin, 1932.

Thomson, G. H., *The Factoral Analysis of Human Ability*, London: University of London Press, 1939.

Thomson, G. H., *The Trend of National Intelligence: The Galton Lecture*, 1946, London: The Eugenics Society and Hamish Hamilton Medical Books, 1947.

Thomson, G. H., *The Education of an Englishman: An Autobiography*, Edinburgh: Moray House Publications, 1969.

Tonking, H. D., 'The Red Cell Count and Cell Diameter of Kenya Natives', *East African Medical Journal*, 13/2 (1936): 43–9.

Tredgold, A. F., 'Inheritance and Educability', *Eugenics Review*, 13/1 (1921): 339–50.

Tredgold, A. F., *Mental Deficiency (Amentia)*, fifth edition (largely re-written), London: Bailliere, Tindall and Cox, 1929.

Tredgold, A. F., *A Text-Book of Mental Deficiency (Amentia)*, sixth edition, London: Bailliere, Tindall and Cox, 1937.

Trowell, H. C., 'The Medical Training of Africans', *East African Medical Journal*, 11/11 (1935): 338–53.

Vint, F. W., 'Gall Stones in a Native Child', *Kenya and East African Medical Journal*, 5/2 (1928): 59–61.

Vint, F. W., 'One Year's Post-Mortem Work on Natives of East Africa', *Kenya and East African Medical Journal*, 5/12 (1929): 383–93.

Vint, F. W., 'Cirrhosis of the Liver of the East African Native', *Kenya and East African Medical Journal*, 7/12 (1931): 349–75.

Vint, F. W., 'Notes on the Pathology of Syphilis', *Kenya and East African Medical Journal*, 8/4 (1931): 94–101.

Vint, F. W., 'A Preliminary Note on the Cell Content of the Prefrontal Cortex of the East African Native', *East African Medical Journal*, 9/2 (1932): 30–55.

Vint, F. W., 'The Brain of the Kenya Native', *Journal of Anatomy*, 68/2 (1934): 216–23.

Vint, F. W., 'Post-Mortem Findings in the Natives of Kenya', *East African Medical Journal*, 13/11 (1937): 332–44.

Vint, F. W., 'The Measurement of Red Blood Corpuscles', *East African Medical Journal*, 16/8 (1939): 295–307.

Vint, F. W., 'Some Recent Researches on the Spleen and Their Possible Relationship to Blackwater Fever', *East African Medical Journal*, 18/6 (1941): 162–73.

Vint, F. W., 'The Pathology of Plague', *East African Medical Journal*, 19/1 (1942): 9–14.

Vint, F. W., 'Solar Rays. Fact or Fiction?', *East African Medical Journal*, 21/8 (1944): 227–39.

von Economo, Constanin, *The Cyroarchitectonics of the Human Cerebral Cortex*, London: Oxford University Press, 1929.

Walter, A., 'Brain and Mind', *Eugenics Review*, 26/4 (1934): 311–12.

Walter, A., 'Climate and White Settlement in the East African Highlands', *East African Medical Journal*, 11/7 (1934): 210–25.

Watson, W., 'A Case of Sexual Perversion in an African Male', *East African Medical Journal*, 20/10 (1943): 354.

Westermann, D., *The African To-Day*, London: Oxford University Press, 1934.

Whetham, W. C. D. and C. D., 'The Influence of Race in History' in *Problems in Eugenics – Papers Communicated to the First International Eugenics Congress*, pp. 237–46, London: Eugenics Education Society, 1912.

Secondary literature

Adams, M. B., 'Towards a Comparative History of Eugenics' in *The Wellborn Science. Eugenics in Germany, France, Brazil, and Russia*, edited by Adams, M. B., pp. 217–31. Oxford: Oxford University Press, 1990.

Adams, M. B. (ed.), *The Wellborn Science. Eugenics in Germany, France, Brazil, and Russia*, Oxford: Oxford University Press, 1990.

Adebola, A. S., 'The London Connection: A Factor in the Survival of the Kikuyu Independent Schools Movement, 1929–39', *Journal of African Studies*, 10/1 (1983): 14–23.

Anderson, D. M., 'Depression, Dustbowl, Demography, and Drought: The Colonial State and Soil Conservation in East Africa during the 1930s', *African Affairs*, 83/332 (1984): 321–43.

Anderson, D. M., 'Policing, Prosecution and the Law in Colonial Kenya, c. 1905–1939', in *Policing the Empire. Government, Authority and Control, 1830–1940*, edited by Anderson, D. M. and Killingray, D., pp. 183–200, Manchester: Manchester University Press, 1991.

Anderson, D. M., 'Black Mischief: Crime, Protest and Resistance in Colonial Kenya', *Historical Journal*, 36/4 (1993): 851–77.

Anderson, D. M., 'The "Crisis of Capitalism" and Kenya's Social History: A Comment', *African Affairs*, 92 (1993): 285–90.

Anderson, D. M., *Histories of the Hanged: Britain's Dirty War and the End of Empire*, London: Weidenfeld and Nicolson, 2005.

Anderson, D. M. and D. Killingray, 'Consent, Coercion and Colonial Control: Policing the Empire, 1830–1940' in *Policing the Empire. Government, Authority and Control, 1830–1940*, edited by Anderson, D. M. and Killingray, D., pp. 1–15, Manchester: Manchester University Press, 1991.

Arnold, D. (ed.), *Imperial Medicine and Indigenous Societies*, Manchester: Manchester University Press, 1988.

Arrighi, G., 'The Political Economy of Rhodesia' in *Essays on the Political Economy of Africa*, edited by Arrighi, G. and Saul, J. S., pp. 336–77, New York: Monthly Review Press, 1973.

Askwith, T., *The Story of Kenya's Progress*, Nairobi: The Eagle Press, 1953.

Askwith, T., *From Mau Mau to Harambee: Memoirs and Memoranda of Colonial Kenya*, Cambridge: African Studies Centre, 1995.

Banton, M., *The Idea of Race*, London: Tavistock Publications, 1977.

Banton, M., *Racial Theories*, Cambridge: Cambridge University Press, 1987.

Banton, M., 'Galton's Conception of Race in Historical Perspective' in *Sir Francis Galton, FRS. The Legacy of His Ideas. Proceedings of the Twenty-Eighth Annual Symposium of the Galton Institute, London*, edited by Keynes, M., pp. 170–9, London: Macmillan, 1993.

Barkan, E., *The Retreat of Scientific Racism: Changing Concepts of Race in Britain and the United States between the World Wars*, Cambridge: Cambridge University Press, 1992.

Barker, D., 'How to Curb the Unfit: The Feeble-Minded in Edwardian Britain', *Oxford Review of Education*, 9/3 (1983): 197–211.

Beck, A., *A History of the British Medical Administration of East Africa, 1900–1950*, Cambridge, Massachusetts: Harvard University Press, 1970.

Beinart, W. and Dubow, S. (eds), *Segregation and Apartheid in Twentieth-Century South Africa*, London: Routledge, 1995.

Bennett, G., 'Imperial Paternalism: The Representation of African Interests in the Kenya Legislative Council' in *Essays in Imperial Government Presented to Margery Perham*, edited by Robinson, K. and Madden, F., pp. 141–69, Oxford: Basil Blackwell, 1963.

Bennett, G., *Kenya, A Political History: The Colonial Period*, London: Oxford University Press, 1963.

Bennett, G., 'Settlers and Politics in Kenya up to 1945' in *Oxford History of East Africa*, Harlow, V. and Chilver, E. M. (eds), pp. 265–332, Oxford: Oxford University Press, 1965.

Bennett, J. H. (ed.), *Natural Selection, Heredity, and Eugenics. Including Selected Correspondence of R. A. Fisher with Leonard Darwin and Others*, Oxford: Clarendon Press, 1983.

Berman, B., *Control and Crisis in Colonial Kenya, The Dialectic of Domination*, London: James Currey, 1990.

Berman, B. and Lonsdale, J., *Unhappy Valley, Vol. 1: State and Class, 2: Violence and Ethnicity*, Oxford: James Curry, 1992.

Berry, V. (ed.), *The Culwick Papers, 1934–1944: Population, Food and Health in Colonial Tanganyika*, London: Academy Books, 1994.

Berry, V. and Petty, C. (eds), *The Nyasaland Survey Papers, 1938–1943: Agriculture, Food and Health*, London: Academy Books, 1992.

Billig, M., 'The Origins of Race Psychology', *Patterns of Prejudice*, 16/3 (1982): 4.

Blacker, C. P., 'The Press and Eugenics: A Review of Reviews', *Eugenics Review*, 26/3 (1934): 207–12.

Blacker, C. P., *Eugenics. Galton and After*, London: Duckworth and Co., 1952.

Blacker, C. P., 'Galton's View on Race' in *Eugenics. Galton and After*, pp. 323–8. London: Duckworth, 1952.

Blanch, M., 'Imperialism, Nationalism and Organized Youth' in *Working Class Culture: Studies in History and Theory*, edited by Clark, J., Critcher, C. and Johnson, R., pp. 103–20, London: Hutchinson, 1979.

Blixen, Karen, *Out of Africa*, London: Penguin, 1954.

Bowler, P. J., *The Non-Darwinian Revolution: Reinterpreting a Historical Myth*, Baltimore and London: The Johns Hopkins University Press, 1992.

Breman, J. (ed.), *Imperial Monkey Business. Racial Supremacy in Social Darwinist Theory and Colonial Practice*, Amsterdam: VU University Press, 1990.

Brett, E. A., *Colonialism and Underdevelopment in East Africa: The Politics of Economic Change, 1919–1939*, London: Heinemann, 1979.

Callan, H. and Ardener, S. (eds), *The Incorporated Wife*, London: Croom Helm, 1984.

Carman, J. A., *A Medical History of the Colony and Protectorate of Kenya. A Personal Memoir*, London: Rex Collings, 1976.

Carnegie, V. M., *Kenyan Farm Diary*, Edinburgh: William Blackwood and Sons, 1930.

Childs, Donald J., *Modernism and Eugenics: Woolf, Yeats and the Culture of Degeneration*, Cambridge: Cambridge University Press, 2001

Clayton, A. and Savage, D., *Government and Labour in Kenya, 1895–1963*, London: Frank Cass, 1974.

Cleminson, R., 'Eugenics by Name or by Nature? The Spanish Anarchist Sex Reform of the 1930s', *History of European Ideas*, 18/5 (1994): 729–40.

Conklin, A. L., *A Mission to Civilize. The Republican Idea of Empire in France and West Africa, 1895 1930*, Stanford, California: Stanford University Press, 1997.

Constantine, Stephen, *The Making of British Colonial Development Policy, 1914–1940*, London: Frank Cass, 1984.

Darwin, L., 'Eugenics and Imperial Development', *Eugenics Review*, 11/3 (1919): 124–35.

Dikötter, F., 'Race Culture: Recent Perspectives on the History of Eugenics', *American Historical Review*, 103/2 (1998): 467–78.

Dilley, M. R., *British Policy in Kenya Colony*, New York: Thomas Nelson and Sons, 1937.

Dowbiggin, I. R., *Keeping America Sane: Psychiatry and Eugenics in the United States and Canada, 1880–1940*, Ithaca: Cornell University Press, 1997.

Drouard, A., 'Eugenics in France and Scandinavia: Two Case Studies' in *Essays in the History of Eugenics*, edited by Peel, R., pp. 173–207, London: The Galton Institute, 1998.

Dubow, S., 'Race, Civilisation and Culture: The Elaboration of Segregationist Discourse in the Inter-War Years' in *The Politics of Race, Class and Nationalism in Twentieth Century South Africa*, edited by Marks, S. and Trapido, S., pp. 71–94, London: Longman, 1987.

Dubow, S., *Scientific Racism in Modern South Africa*, Cambridge: Cambridge University Press, 1995.

Dubow, S. (ed.), *Science and Society in Southern Africa*, Manchester: Manchester University Press, 2000.

Duder, C. J. D., '"Men of the Officer Class": the Participants in the 1919 Soldier Settlement Scheme in Kenya', *African Affairs*, 92 (1993): 69–87.

Duder, C. J. D. and Youé, C. P., 'Paice's Place: Race and Politics in Nanyuki District, Kenya, in the 1930s', *African Affairs*, 93 (1994): 253–78.

East Africa Women's League, *They Made It Their Home*, Nairobi: East African Standard Ltd, 1962.

Elkins, C., *Britain's Gulag: The Brutal End of Empire in Kenya*, London, Jonathan Cape, 2005.

Engels, D. and Marks, S. (eds), *Contesting Colonial Hegemony: State and Society in Africa and India*, London: British Academic Press, 1994.

Ernst, W., 'The European Insane in British India, 1800–1858: A Case Study in Psychiatry and Colonial Rule' in *Imperial Medicine and Indigenous Societies*, edited by Arnold, D., pp. 27–44, Manchester: Manchester University Press, 1988.

Farrall, L., 'The History of Eugenics: a Bibliographical Review', *Annals of Science*, 36 (1979): 111–23.

Farrall, L., *The Origins and Growth of the English Eugenics Movement, 1865–1925*, New York: Garland Publishing, 1985.

Foran, W. R., *The Kenya Police 1887–1960*, London: Robert Hale, 1962.

Forrest, D. W., *Francis Galton. The Life and Work of a Victorian Genius*, Elek, 1974.

Freeden, M., 'Eugenics and Progressives: A Study in Ideological Affinity', *Historical Journal*, 22/3 (1979): 645–71.

Freeden, M., 'Eugenics and Ideology', *Historical Journal*, 26/4 (1983): 959–62.

Fryer, P., *Staying Power, the History of Black People in Britain*, London: Pluto, 1984.

Furedi, F., 'The African Crowd in Nairobi: Popular Movements and Elite Politics', *Journal of African History*, 14/2 (1973): 275–90.

Furley, O. W. and Watson, T., *A History of Education in East Africa*, New York: NOK Publishers, 1978.

Ghai, Y. P. and McAuslan, J. P. W. B., *Public Law and Political Change in Kenya. A Study of the Legal Framework of Government from Colonial Times to the Present*, London: Oxford University Press, 1970.

Giliomee, H., 'Processes in the Development of the South African Frontier' in *The Frontier in History*, edited by Lamar, H. and Thompson, L., pp. 76–119, New Haven: Yale University Press, 1981.

Gillett, M., *Tribute to Pioneers. Mary Gillett's Index of Many of the Pioneers of East Africa*, Oxford: J. M. Considine (published privately), 1986.

Gillis, J., 'The Evolution of Juvenile Delinquency in England', *Past and Present*, 67 (1975): 96–126.

Gilroy, P., *'There Ain't No Black in the Union Jack': The Cultural Politics of 'Race' and Nation*, London: Routledge, 1987.

Gould, S. J., *The Mismeasure of Man*, London: Penguin Books, 1997.

Graham, L. R., 'Science and Values: The Eugenics Movement in Germany and Russia in the 1920s', *American Historical Review*, 83/5 (1978): 1135–64.

Gregory, J. R., *Under the Sun (A Memoir of Dr R. W. Burkitt of Kenya)*, Nairobi, privately published, 1951.

Gregory, R. G., *Sidney Webb and East Africa: Labour's Experiment with the Doctrine of Native Paramountcy*, Berkeley: University of California Press, 1962.

Hall, L., 'Women, Feminism and Eugenics' in *Essays in the History of Eugenics*, edited by Peel, R., pp. 36–51, London: The Galton Institute, 1998.

Halliday, R. J., 'Social Darwinism: A Definition', *Victorian Studies*, 14/4 (1971): 389–405.

Harlow, V. and Chilver, E. M. (eds), *Oxford History of East Africa*, Vol. II., Oxford: Oxford University Press, 1965.

Halliday, John, *Darwinism, Biology and Race*, Working Paper no. 49, University of Warwick, 1990.

Hartz, L., *The Founding of New Societies*, New York: Harcourt Brace and World, 1964.

Hetherington, P., *British Paternalism and Africa, 1920–1940*, London: Frank Cass, 1978.

Hetherington. P., 'Explaining the Crisis of Capitalism in Kenya', *African Affairs*, 92 (1993): 89–103.

Hobsbawm, E., 'The Fabians Reconsidered' in *Labouring Men: Studies in the History of Labour*, London: Weidenfeld and Nicolson, 1968.

Humphries, Stephen, *Hooligans or Rebels?, An Oral History of Working Class Childhood and Youth, 1889–1939*, Oxford: Blackwell, 1981.

Huxley, E., *White Man's Country: Lord Delamere and the Making of Kenya*, Volumes I and II, London: Macmillan, 1935.

Huxley, E., *No Easy Way. A History of the Kenya Farmers' Association and Unga Limited*, Nairobi: The East African Standard, 1957.

Huxley, E. (ed.), *Nellie. Letters from Africa*, London: Weidenfeld and Nicolson, 1973.

Huxley, E. and Perham, M., *Race and Politics in Kenya*, London: Faber and Faber, 1944.

Hyslop, J., 'White Working-Class Women and the Invention of Apartheid: "Purified" Afrikaner Nationalist Agitation for Legislation against "Mixed Marriages", 1934–9', *Journal of African History*, 36/1 (1995): 57–81.

Iliffe, J., *East African Doctors*, Cambridge: Cambridge University Press, 1998.

Jeffries, C., *The Colonial Office*, London: Allen and Unwin, 1956.

Jones, G., *Social Darwinism in English Thought: The Interaction between Biological and Social Theories*, Brighton: Harvester Press, 1980.

Jones, G., 'Eugenics and Social Policy between the Wars', *Historical Journal*, 25/3 (1982): 717–28.

Jones, G., *Social Hygiene in Twentieth Century Britain*, London: Croom Helm, 1986.

Jones, G., 'The Theoretical Foundations of Eugenics' in *Essays in the History of Eugenics*, edited by Peel, R., pp. 1–19, London: The Galton Institute, 1998.

Kanogo, T., 'Rift Valley Squatters and Mau Mau', *Kenya Historical Review*, 5 (1977): 243–53.

Kennedy, D., *Islands of White: Settler Society and Culture in Kenya and Southern Rhodesia, 1890–1939*, Durham: Duke University Press, 1987.

Kevles, D., 'Genetics in the United States and Great Britain 1890–1930: A Review with Speculations' in *Biology, Medicine and Society 1840–1940*, edited by Webster, C., pp. 193–215, Cambridge: Cambridge University Press, 1981.

Kevles, D., *In the Name of Eugenics: Genetics and the Uses of Human Heredity*, Cambridge, Mass.: Harvard University Press, 1995.

Kevles, D., 'Eugenics in North America' in *Essays in the History of Eugenics*, edited by Peel, R., pp. 208–26. London: The Galton Institute, 1998.

Kiernan, V., *The Lords of Human Kind. European Attitudes to Other Cultures in the Imperial Age*, London: Serif, 1995.

Killingray, D. and Ellis, S., 'Introduction', *African Affairs*, 99/395 (2000): 177–82.

King, K. J., 'The Politics of Agricultural Education for Africans in Kenya' in *Hadith 3*, edited by Ogot, B. A., 142–56. Nairobi: East African Publishing House, 1971.

King, K. J., *Pan-Africanism and Education: A Study of Race Philanthropy and Education in the Southern States of America and East Africa*, Oxford: Clarendon Press, 1971.

Kipkorir, B. E., 'The Functionary in Kenya's Colonial System' in *Biographical Essays on Imperialism and Collaboration in Colonial Kenya*, edited by Kipkorir, B. E., pp. 1–14, Nairobi: Kenya Literature Bureau, 1980.

Kitching, G., *Class and Economic Change in Kenya*, London: Yale University Press, 1980.

Klausen, S., '"For the Sake of Race": Eugenic Discourses of Feeblemindedness and Motherhood in the South African Medical Record, 1903–1926', *Journal of Southern African Studies*, 23/1 (1997): 27–50.

Klausen, S., 'The Race Welfare Society: Eugenics and Birth Control in Johannesburg, 1930–40' in *Science and Society in Southern Africa*, edited by Dubow, S., pp. 164–87, Manchester: Manchester University Press, 2000.

Klausen, S. M., *Race, Maternity and the Politics of Birth Control in South Africa, 1910–39*, Basingstoke: Macmillan, 2004.

Kline, P., 'Ninety Years of Psychometrics' in *Essays in the History of Eugenics*, edited by Peel, R., pp. 128–55, London: The Galton Institute, 1998.

Kohn, M., *The Race Gallery*, London: Vintage, 1995.

Kopytoff, I. (ed.), *The African Frontier: The Reproduction of Traditional African Societies*, Bloomington: Indiana University Press, 1987.

Lamar, H. and Thompson, L. (eds), *The Frontier in History: North America and Southern Africa Compared*, New Haven: Yale University Press, 1981.

Langford, C., 'The Eugenics Society and the Development of Demography in Britain: The International Population Union, the British Population Society and the Population Investigation Committee' in *Essays in the History of Eugenics*, edited by Peel, R., pp. 81–111, London: The Galton Institute, 1998.

Lewis, J., *Empire State-Building: War and Welfare in Kenya, 1925–52*, Oxford: James Currey, 2000.

Leys, C., *Underdevelopment in Kenya: The Political Economy of Neo-colonialism, 1964–1971*, London: Heinemann, 1975.

Lonsdale, J. M., 'Some Origins of Nationalism in East Africa', *Journal of African History*, 9/1 (1968): 119–46.

Lonsdale, J. M., 'European Attitudes and African Pressures: Missions and Government in Kenya between the Wars' in *Hadith 2*, edited by Ogot, B. A., pp. 229–42, Nairobi: East African Publishing House, 1970.

Lonsdale, J. M., 'Mau Maus of the Mind: Making Mau Mau and Remaking Kenya', *Journal of African History*, 31 (1990): 393–422.

Lonsdale, J. M., 'The Moral Economy of Mau Mau' in *Unhappy Valley: Conflict in Kenya and Africa, Book Two: Violence and Ethnicity*, edited by Berman, B. and Lonsdale, J. M., pp. 265–504, Oxford: James Currey, 1992.

Love, R., '"Alice in Eugenics-Land": Feminism and Eugenics in the Scientific Careers of Alice Lee and Ethel Elderton', *Annals of Science*, 36 (1936): 145–58.

Low, D. A., 'British East Africa: the Establishment of British Rule' in *Oxford History of East Africa*, edited by Harlow, V. and Chilver, E. M., Oxford: Oxford University Press, 1965.

Low, D. A. and Smith, A. (eds), *Oxford History of East Africa*, Vol. III, Oxford: Clarendon Press, 1976.

Lugumba, S. M. E. and Ssekamwa, J. C., *A History of Education in East Africa (1900–1973)*, Kampala: Kampala Bookshop, 1973.

MacKenzie, D., 'Eugenics in Britain', *Social Studies of Science*, 6 (1976): 499–532.

MacKenzie, D., 'Karl Pearson and the Professional Middle Class', *Annals of Science*, 36 (1979): 125–43.

MacKenzie, D., 'Sociobiologics in Competition: the Biometrician–Mendelian Debate' in *Biology, Medicine and Society 1840–1940*, edited by Webster, C., pp. 243–88, Cambridge: Cambridge University Press, 1981.

MacKenzie, D., *Statistics in Britain, 1865–1930: The Social Construction of Scientific Knowledge*, Edinburgh: Edinburgh University Press, 1981.

Mackenzie, F., *Land, Ecology and Resistance in Kenya, 1880–1952*, Edinburgh: Edinburgh University Press, 1988.

MacKenzie, J. M., *Propaganda and Empire: the Manipulation of British Public Opinion, 1880–1960*, Manchester: Manchester University Press, 1986.

MacKenzie, J. M. (ed.), *Imperialism and Popular Culture*, Manchester: Manchester University Press, 1986.

MacKenzie, J. M., *The Empire of Nature: Hunting, Conservation and British Imperialism*, Manchester: Manchester University Press, 1988.

Malik, K., *The Meaning of Race: Race, History and Culture in Western Society*, Basingstoke: Macmillan Press, 1996.

Malowany, M., 'Unfinished Agendas: Writing the History of Medicine of Sub-Saharan Africa', *African Affairs*, 99/395 (2000): 325–49.

Mangan, J. A., *The Games Ethic and Imperialism. Aspects of the Diffusion of an Ideal*, Harmondsworth: Penguin, 1986.

May, M., 'Innocence and Experience: The Evolution of the Concept of Juvenile Delinquency in the Mid-Nineteenth Century', *Victorian Studies*, 18/1 (1973): 2–29.

Mazumdar, P., 'The Eugenists and the Residuum: The Problem of the Urban Poor', *Bulletin of the History of Medicine*, 54/2 (1980): 204–15.

Mazumdar, P., *Eugenics, Human Genetics and Human Failings. The Eugenics Society, Its Sources and Its Critics in Britain*, London: Routledge, 1992.

McClintock, A., *Imperial Leather. Race, Gender and Sexuality in the Colonial Contest*, New York: Routledge, 1995.

McCulloch, J., *Colonial Psychiatry and 'the African Mind'*, Cambridge: Cambridge University Press, 1995.

Middleton, John, 'Kenya: Administration and Changes in African Life, 1912–45' in *Oxford History of East Africa*, edited by Harlow, V. and Chilver, E. M., pp. 333–92, Oxford: Oxford University Press, 1965.

Morris, H. F. and Read, J. S., *Indirect Rule and the Search for Justice. Essays in East African Legal History*, Oxford: Clarendon Press, 1972.

Mosley, P., *The Settler Economies: Studies in the Economic History of Kenya and Southern Rhodesia, 1900–1963*, Cambridge: Cambridge University Press, 1983.

Mungeam, G. H., *British Rule in Kenya, 1895–1912: The Establishment of Administration in the East Africa Protectorate*, Oxford: Clarendon Press, 1966.

Ndungo, J. B., 'Gituamba and Kikuyu Independency in Church and School' in *Ngano, Studies in Traditional and Modern East African History*, edited by McIntosh, B. G., pp. 131–50, Nairobi: East African Publishing House, 1969.

Norton, B., 'Psychologists and Class' in *Biology, Medicine and Society 1840–1940*, edited by Webster, C., pp. 289–314, Cambridge: Cambridge University Press, 1981.

Nye, R. A., 'The Rise and Fall of the Eugenics Empire: Recent Perspectives on the Impact of Biomedical Thought in Modern Society', *Historical Journal*, 36/3 (1993): 687–700.

Odhiambo, E. S. A., *The Paradox of Collaboration and Other Essays*, Nairobi: Kenya Literature Bureau, 1974.

Ogot, B. A. (ed.), *Hadith 2*, Nairobi: East African Publishing House, 1970.

Ogot, B. A. (ed.), *Hadith 3*, Nairobi: East African Publishing House, 1971.

Orr, J. B., *As I Recall*, London: MacGibbon and Kee Ltd., 1966.

Packard, R. M., *White Plague, Black Labour. Tuberculosis and the Political Economy of Health and Disease in South Africa*, Pietmaritzburg, 1990.

Paice, E., *Lost Lion of Empire: The Life of 'Cape-to-Cairo' Grogan*, London: HarperCollins, 2001.

Parkinson, C., *The Colonial Office from Within, 1909–1945*, London: Faber and Faber, 1947.

Patton, A., *Physicians, Colonial Racism, and Diaspora in West Africa*, Gainesville: University Press of Florida, 1996.

Paul, D. B., 'Eugenics and the Left', *Journal of the History of Ideas*, 45/4 (1984): 567–90.

Pearce, R. D., *The Turning Point in Africa: British Colonial Policy, 1938–1948*, London: Frank Cass, 1982.

Pearson, Roger, *Eugenics and Race*, BCM-Thule, 1966.

Pedersen, S., 'National Bodies, Unspeakable Acts: The Sexual Politics of Colonial Policy-Making', *Journal of Modern History*, 63/4 (1991): 647–80.

Peel, R. A. (ed.), *Marie Stopes, Eugenics and the English Birth Control Movement. Proceedings of a Conference Organised by the Galton Institute*, London, 1996, London: The Galton Institute, 1997.

Peel, R. A. (ed.), *Essays in the History of Eugenics*, London: The Galton Institute, 1998.

Pick, D., *Faces of Degeneration: A European Disorder, c.1848–c.1918*, Cambridge: Cambridge University Press, 1989.

Ray, L. J., 'Eugenics, Mental Deficiency and Fabian Socialism between the Wars', *Oxford Review of Education*, 9/3 (1983): 213–22.

Rich, P., 'The Long Victorian Sunset: Anthropology, Eugenics and Race in Britain, c.1900–48', *Patterns of Prejudice*, 18/3 (1984): 3–17.

Rich, P., *Race and Empire in British Politics*, Cambridge: Cambridge University Press, 1990.

Roberts, A. D., 'The Gold Boom of the 1930s in Eastern Africa', *African Affairs*, 85/341 (1986): 545–62.

Rose, S., 'Scientific Racism and Ideology: The IQ Racket from Galton to Jensen' in *The Political Economy of Science: Ideology of/in the Natural Sciences*, edited by Rose, H. and Rose, S., London: Macmillan, 1976.

Ross, Robert, *Status and Respectability in the Cape Colony, 1750–1870: A Tragedy of Manners*, Cambridge: Cambridge University Press, 1999.

Schilling, D. G., 'Local Native Councils and the Politics of Education in Kenya, 1925–1939', *International Journal of African Historical Studies*, 9/2 (1976): 218–47.

Schneider, W. H., *Quality and Quantity: The Quest for Biological Regeneration in Twentieth-Century France*, Cambridge, Cambridge University Press, 1990.

Searle, G. R., *Eugenics and Politics in Britain, 1900–1914*, Leyden: Noordhalf International Publishing, 1976.

Searle, G. R., 'Eugenics and Politics in Britain in the 1930s', *Annals of Science*, 36 (1979): 159–69.

Searle, G. R., 'Eugenics and Class' in *Biology, Medicine and Society 1840–1940*, edited by Webster, C., pp. 217–42, Cambridge: Cambridge University Press, 1981.

Searle, G. R., *The Quest for National Efficiency. A Study in British Politics and Political Thought, 1899–1914*, London: The Ashfield Press, 1990.

Searle, G. R., 'Eugenics: The Early Years' in *Essays in the History of Eugenics*, edited by Peel, R., pp. 20–35, London: The Galton Institute, 1998.

Semmel, R., *Imperialism and Social Reform, English Social-Imperial Thought 1895–1914*, London: Allen & Unwin, 1960.

Shaw, C., 'Eliminating the Yahoos: Eugenics, Social Darwinism and Five Fabians', *History of Political Thought*, 8/3 (1987): 522–44.

Shaw, C. M., *Colonial Inscriptions. Race, Sex, and Class in Kenya*, Minneapolis: University of Minnesota., 1995.

Simmons, H. G., 'Explaining Social Policy: The English Mental Deficiency Act of 1913', *Journal of Social History*, 11/3 (1978): 387–403.

Soloway, R. A., *Birth Control and the Population Question in England*, Chapel Hill: University of North Carolina Press, 1982.

Soloway, R. A., 'Counting the Degenerates: the Statistics of Race Deterioration in Edwardian England', *Journal of Comparative History*, 17/1 (1982): 137–64.

Soloway, R. A., *Demography and Degeneration: Eugenics and the Declining Birthrate in Twentieth-Century Britain*, Chapel Hill: University of North Carolina Press, 1990.

Soloway, R. A., 'From Mainline to Reform Eugenics – Leonard Darwin and C P Blacker' in *Essays in the History of Eugenics*, edited by Peel, R., pp. 52–80, London: The Galton Institute, 1998.

Sorrenson, M. P. K., *Origins of European Settlement in Kenya*, Nairobi: Oxford University Press, 1968.

Spencer, John, *The Kenya African Union*, London: KPI Limited, 1985.

Stepan, N., *The Idea of Race in Science: Great Britain, 1800–1960*, London: Macmillan Press, 1982.

Stepan, N., 'Eugenics in Brazil' in *The Wellborn Science. Eugenics in Germany, France, Brazil, and Russia*, edited by Adams, M. B., Oxford: Oxford University Press, 1990.

Stepan, N., *'The Hour of Eugenics'. Race, Gender, and Nation in Latin America*, Ithaca: Cornell University Press, 1991.

Strayer, R. W. in conjunction with Jo Murray, 'The CMS and Female Circumcision' in *The Making of Mission Communities in East Africa: Anglicans and Africans in Colonial Kenya, 1875–1935*, edited by Strayer, R. W., pp. 136–55, London: Heinemann, 1978.

Thomas, L. M., 'Imperial Concerns and "Women's Affairs": State efforts to regulate clitoridectomy and eradicate abortion in Meru, Kenya, c. 1910–1950', *Journal of African History*, 39/1 (1998): 121–45.

Thomson, M., *The Problem of Mental Deficiency. Eugenics, Democracy, and Social Policy in Britain c. 1870–1959*, Oxford: Clarendon Press, 1998.

Throup, D., *Economic and Social Origins of Mau Mau 1945–1953*, London: James Currey, 1987.

Timson, J., 'Human Genetics' in *Essays in the History of Eugenics*, edited by Peel, R., pp. 112–27, London: The Galton Institute, 1998.

van Zwanenberg, R. M. A., *Colonial Capitalism and Labour in Kenya, 1919–1939*, Nairobi: East African Literature Bureau, 1975.

Vaughan, M., 'Idioms of Madness: Zomba Lunatic Asylum, Nyasaland, in the Colonial Period', *Journal of Southern African Studies*, 9/2 (1983): 218–38.

Vaughan, M., *Curing Their Ills – Colonial Power and African Illness*, Stanford: Stanford University Press, 1991.

Webster, C., 'Introduction' in *Biology, Medicine and Society 1840–1940*, edited by Webster, C., pp. 1–13, Cambridge: Cambridge University Press, 1981.

Willis, J., 'Thieves, Drunkards and Vagrants: Defining Crime in Colonial Mombasa, 1902–32' in *Policing the Empire: Government, Authority and Control, 1830–1940*, edited by Anderson, D. M. and Killingray, D., pp. 219–35, Manchester: Manchester University Press, 1991.

Wintle, M. (ed.), *Culture and Identity in Europe*, Aldershot: Avebury, 1996.

Wrigley, C. C., 'Kenya: the Patterns of Economic Life, 1902–1945' in *Oxford History of East Africa*, edited by Harlow, V. and Chilver, E. M., pp. 209–64, Oxford: Oxford University Press, 1965.

Wylie, D., 'Confrontation Over Kenya: The Colonial Office and Its Critics, 1918–1940', *Journal of African History*, 18/3 (1977): 427–47.

Unpublished theses

Beuschel, Gail E., 'Shutting Africans Away: Lunacy, Race and Social Order in Colonial Kenya, 1910–1963', PhD thesis, University of London, 2001.

Bin Harun, H., 'Medicine and Imperialism: a Study of the British Colonial Medical Establishment, Health Policy and Medical Research in the Malay Peninsula, 1786–1918', PhD thesis, University of London, 1988.

Burton, A., 'Wahuni (The Undesirables): African Urbanisation, Crime and Colonial Disorder in Dar es Salaam, 1919–1961', PhD thesis, University of London, 2000.

Lyons, A. P., 'The Question of Race in Anthropology from the Time of Johann Friedrich Blumenbach to that of Franz Boas', PhD thesis, Oxford University, 1974.

Mungai, J. M., 'Dendritic Patterns in the Somatic Sensory Cortex', PhD thesis, University of London, 1966.

Redley, M. G., 'The Politics of a Predicament: The White Community in Kenya 1918–1932', PhD thesis, Cambridge University, 1976.

Tilley, H., 'Africa as a Living Laboratory: The Colonial Research Survey and the British Colonial Empire', PhD thesis, University of Oxford, 2001.

Worboys, M., 'Science and British Colonial Imperialism, 1895–1940', PhD thesis, University of Sussex, 1980.

Young, T., 'The Eugenics Movement and the Eugenic Idea in Britain, 1900–1914', PhD thesis, University of London, 1980.

EU authorised representative for GPSR:
Easy Access System Europe, Mustamäe tee 50,
10621 Tallinn, Estonia
gpsr.requests@easproject.com

www.ingramcontent.com/pod-product-compliance
Ingram Content Group UK Ltd.
Pitfield, Milton Keynes, MK11 3LW, UK
UKHW021909060726
6981IPUK00003B/46